SIGNS IN AMERICA'S AUTO AGE

SIGNS

AMERICAN LAND AND LIFE SERIES

Edited by Wayne Franklin

in America's Auto Age

SIGNATURES OF LANDSCAPE AND PLACE

JOHN A. JAKLE

AND

KEITH A. SCULLE

FOREWORD BY

WAYNE FRANKLIN

University of Iowa Press

Iowa City

University of Iowa Press, Iowa City 52242

Printed in the United States of America
Design by Richard Hendel
http://www.uiowa.edu/uiowapress

Unless otherwise noted, all photographs are by the authors.

The publication of this book was generously supported by the University of Iowa Foundation.

Printed on acid-free paper

Library of Congress Cataloging-in-Publication Data
Signs in America's auto age: signatures of landscape and place / by John A. Jakle and Keith A. Sculle.
p. cm. — (American land and life series)
Includes bibliographical references and index.
ISBN 0-87745-889-8 (cloth), ISBN 0-87745-890-1 (pbk.)
1. Signs and signboards—United States. 2. Advertising, Outdoor—United States. 3. Social interaction—United States. 4. Landscape—United States. 5. Cities and towns—United States. I. Title. II. Series.
HF5841.J35 2004
659.13'42'0973—dc22 2003063364

04 05 06 07 08 C 5 4 3 2 1
04 05 06 07 08 P 5 4 3 2 1

To Cindy and Tracey

ORLD'S HIGHEST STANDARD OF LIV

There's no w
like the
American

RLD'S SHORTEST WORKING HO

There's no w
like the
American

ORLD'S HIGHEST WAG

here's no way
like the
nerican Way

Contents

Foreword by *Wayne Franklin*, ix

Preface and Acknowledgments, xvii

Introduction, xxi

PART ONE: COMMERCIAL SIGNS

1 Signs Downtown, 3

2 Signs on Main Street, 18

3 Roadside Signs, 32

PART TWO: SIGNING PUBLIC PLACES

4 Traffic Signs, 55

5 Signs and Community, 73

PART THREE: SIGNING PERSONAL SPACE

6 Territorial Markers and Signs of Personal Identity, 95

PART FOUR: SIGN AESTHETICS

7 Signs and Landscape Visualization, 117

8 Sign Regulation, 144

Epilogue, 167

Notes, 171

Bibliography, 193

Index, 213

DIVIDED HIGHWAY
ROAD NARROWS
MEN WORKING
HILL

Foreword *Wayne Franklin*

Signage so pervades the modern American scene that it is hard to imagine how this landscape looked when the sign did not intervene between eye and place. In fact, one has to go very far back in order to find the "clean" landscape. Centuries before the insertion of the signboards and traffic devices and other objects that interest landscape historians John Jakle and Keith Sculle in the modern era, the European eye seemed to crave some interposition of cultural signs where familiar objects were all too literally few and far between. In his innovative historical novel *LaSalle*, published in 1986, John Vernon gives to his doomed title character a chance to fantasize an overlay of French landscape on the rolling grasslands of the new continent's heart:

> France must have been like this before the Romans came; the Romans drove out the barbarians and became the Gauls, building villages and farms in valleys such as this one. Gazing down at this valley, I imagined hedgerows, walls, villages, footpaths, mills, fairs, churches, commerce, sheep and cows at pasturage, cultivated fields, estates and chateaus, all the appurtenances of a thriving civilization; here a wedding feast, there some women washing clothes on a creek bank; smoke rising from the blacksmith's shop, and barges carrying livestock down the river. Then all dissolved and rose like a mist; the buffaloes returned, along with the unmowed grass and groves of trees, and the entire valley unfurled like a very carpet, untrodden by either coarse boots or slippers.

Vernon is right: the earliest descriptions of the American landscape by European explorers and settlers soon become meditations on signs and their absence from the new lands. When, beginning his famous history *Of Plymouth Plantation* in 1630, William Bradford sought to recall what Cape Cod had looked like on the arrival of his company a decade earlier, he was at a loss for words. He could recover its original appearance only by remembering the English landscapes the settlers had left and then *erasing* from such older memories the signs of habitation that now made his actually hostile homeland seem familiar, even welcoming, in retrospect. The residue, what was left after all those erasures, was America. In a formula that would long persist in American landscape description, Bradford thus emphasized what the new place lacked: as the Pilgrims first gazed upon it, they found "no friends to welcome them nor inns to entertain their weatherbeaten bodies; no houses or much less towns to repair to, to seek for succour." Such absent things were, of course, material realities first and foremost. Had they been present on the Cape in 1620, they would have offered precisely the sorts of accommodation the isolated adventurers needed at that time. Indeed, we know from an earlier account by Bradford that among the last pleasantries the Pilgrims associated with England was the fact that they had been "kindly entertained and courteously used by divers friends there dwelling" just before boarding the *Mayflower* — a fact that gives real emotional freight to what America, in his eyes, lacked. But the absent things were also, despite the particularity of this one memory, *signs* as well. They represented not just actual events but the whole recollected order of the other world and therefore expressed the perceived disorder — the alien nature — of the landscapes all-too-palpably *there* before the settlers arrived. America was a domain bereft not only of the things that made any landscape habitable but also of the signs of better things to come in Plymouth.

From the moment the Pilgrims chose their intended site on the mainland at Plymouth and then stepped off their vessel, they began strewing the land with necessary things that were also, always, signs. A house in old Yorkshire, astride the road leading from Bradford's native village of Austerfield to the nearby manor house at Scrooby, where his friend and fellow immigrant William Brewster once lived, was first of all a house — one structure among many that grew from and extended but did not necessarily represent the social ordering of the land in quite the way that houses came to represent ideas in America. In the new world settlements, so frail was the presence of substantial order that any item of European design quickly came to be a conscious sign as well as a useful object. For generations, the colonists read the land in terms of the density or scarcity of imported artifacts. Settling the land involved not only organizing it for the sake of its economic exploitation and therefore the accommodation of the settlers, but also determining and proclaiming the meaning of the continent — something for which the signs of actual human use proved quite effective.

All of this layering of meaning in the landscape was enriched even more by the old European tendency of seeing the natural world as a vast interconnected system of signs. Plants were "read" by their form, location, or habits as having certain uses in the domestic economy, for food or home manufactures or especially for medicinal purposes. Nature was not viewed solely as a material realm in which things were simply themselves. One thing revealed its covert relationship to others by "signatures" that it was thought to bear. This imported theory of signatures allowed American immigrants to infer that the unfamiliar sights greeting their eyes had meaning both as individual items and as assemblages. The details might be new, that is, but the form was the same.

The theory of signatures did not lessen, to be sure, the immediate impact of all the newness that Europeans found on landing in the Americas. Slowly they appropriated much of what they found by borrowing ideas, recipes, and names from old world analogues. They looked, first of all, for American things that were versions of European counterparts. Both pine trees and sparrows, to take two obvious instances, were already familiar to them. Yet often this memorial recognition of objects in the landscape around them hardly controlled or even noticeably quieted the clamoring newness. In New Netherland, the Dutch are said (in not wholly reliable but nonetheless revealing legends) to have named as many of the newly encountered species of fish as they could after their analogues in the old country. But so rich were the new land's waterborne resources that many species were completely new and names for them were wanting. Hence began the quasi-humorous custom of *numbering* the subsequent discoveries — the "eleventh" (shad), the "twelfth" (striped bass), and the "thirteenth" (drum). The custom may have originated with a convenient pun: the name of the shad in the Netherlands (the *elft*) was close to the word for eleven (*elf*). But the New Netherland names for the bass and drum were coinages whose humorous nature showed the challenges placed on old world knowledge and language by new world realities. These new terms for fish that were of much commercial interest along the American coast suggest the degree to which self-conscious verbalism might come into play in the fields of new world abundance.

■ It is a long way, in some senses, from these early efforts at naming and reading the fresh American scene to the layered communications systems that John Jakle and Keith Sculle dissect with illuminating authority in the pages that follow. In speaking of signs in the manner employed above, I am interpreting the concept in what Jakle and Sculle rightly call "semiotic" terms. Semiotics, although as a field it has roots in linguistics, is interested in much more than verbal signs. To a semiotician, anything that is thought to represent something else, often covertly rather than overtly, is a sign. Although Jakle and Sculle are adept at borrowing a number of useful hints from semiotics, what interests them as much as the abstract *sign* is the material *signboard*, a quite real surface made to bear explicit, frequently textual and extra-textual messages. On this subject they are masterful. They divide the landscape into both its functioning zones (downtown, the roadside, private property) and its jurisdictional regimes (commerce, the state, the individual). On all these topics, with a talent for analysis and description alike, they offer compelling images and commentary. Their sources are splendid: the landscape itself, of course, but also the rich archives of trade and professional magazines like *Signs of the Times* and the *Highway Research Record*, technical publications (like two volumes issued by the Institute of Traffic Engineers, *Traffic Devices* and *Traffic Engineering Handbook*), and the growing body of roadside monographs and pictorial treatments published by architects, popular culture scholars, collectors, and enthusiasts. While their analysis of textualized signboards is illuminating, they also delve into the nonverbal aspects of roadside communication — from devices for controlling vehicular traffic to what Robert Venturi and his colleagues dubbed the "duck" (from an outlet on a Long Island duck farm) — a building that is shaped like the items sold in it. They round out their discussion with a consideration of graffiti, territorialism, and personal space.

Moreover, Jakle and Sculle situate their whole treatment of the nature of signs in a sophisticated behavioral context. Most simply, signs identify places in terms of their nature (e.g., CITY PARK). But they also name or otherwise indicate the kinds of activities which are permitted or forbidden in a given place (PICNIC AREA or NO SWIMMING, for both of which there are now widely used symbol signboards as well). In certain contexts — not all benign, obviously — signs discriminate among potential users of a given space (MEMBERS ONLY or RESIDENTS ONLY — but also WHITES ONLY). In commercial contexts, signs may range from the simplest of imperatives (EAT) to the most bizarre of persuasive come-ons (FREE ICE WATER: WALL DRUG, WALL, S.D., 553 MILES). Most of these kinds of signs have their nonverbal accompaniments or equivalents: not only the modern international symbology of the tourist landscape but also the older markers, such as the color red for the building exit sign, big pictures of food items like hot dogs or hamburgers or donuts, security gates and kiosks at residential compounds, Jersey barriers at restricted locales, and the like. All in all, as Jakle and Sculle argue, signs not only communicate their identity of the places they name but also delineate their boundaries, both spatial and behavioral.

■ It is this second insight that gives *Signs in America's Auto Age: Signatures of Landscape and Place* its deepest appeal. I recall a particularly intriguing example of a private sign that illustrates many of Jakle and Sculle's themes. Some twenty-five years ago I was driving through South Carolina visiting old houses and historic sites. One of the houses I most wanted to locate and see was the early eighteenth-century mansion known as the Mulberry. This structure, erected in 1714 along the west branch of the Cooper River upstream from Charleston, is of a lovely composite design. French, Ducth, and English precedents have all been cited by those who have studied it. It was built to serve as the center of a prospective silk cultivation experiment (hence its name and the symbol of a mulberry branch inside a horseshoe that adorns the pediment of the front porch). One of the pictures I had seen of the building showed it in a run-down condition at the start of the twentieth century, but I was determined to locate the site and, if possible, explore the building's current condition. The old composite *American Guide* of 1949, based on the individual state volumes produced by the Federal Writers Project of the WPA, keyed the building to a mileage point along U.S. Route 52 where a winding mile-long private drive turned off to the right. Since the publication of that volume, U.S. 52 apparently had been relocated, however, so finding the intersection in question proved vexing. Eventually, by a process of elimination, I came across the superceded route and discovered a suitably ancient-looking historical marker standing at the public end of the narrow, shady lane that must have been the private road mentioned in my 1949 source. The guidebook claimed that "Mulberry Castle" was "open in season," but there was no sign that such was the case any longer. The property seemed to have receded once more into the private domain. I inched past the historical marker and turned into the lane, determined not to retreat now that I had found the place. Just on the right, however, I was stopped in my tracks by the most curious sign I have ever encountered in several decades of desultory touring both in the United States and abroad. As near as I can remember it, its wording ran as follows:

> DO NOT ENTER THIS PROPERTY LOOKING FOR YOUR DOG. IF WE FIND YOUR DOG WE WILL CONTACT YOU. VIOLATORS WILL BE PROSECUTED.

I paused to deconstruct this curious warning. To begin with, it was so personal. It assumed that *any* reader of the sign not only *owned* a dog but had *lost* it and was out searching for it. While not denying that any of those putative dogs just might be on the extensive property, and indeed holding out the halfway courteous promise that any found there would be returned, the sign absolutely forbade dog owners in search of their stray hounds, spaniels, or mutts from proceeding.

Exactly what, I wondered, was going on? For a moment I felt rather the way I had when I first toured the South even before 1978. At that time, I happened to stop at a gas station in Hope, Arkansas (now no doubt posted with signs declaring it THE BOYHOOD HOME OF WILLIAM JEFFERSON CLINTON), and waited while the attendant filled the tank and cleaned the windshield. When he had finished and been paid, he flashed a big smile and said, "Ya'll come back ag'in now, ya hear?" Having never heard that formulaic southern farewell before, I felt compelled to answer it in the terms that it seemed to set in motion. So, in my halting northern dialect, I said something like, "Well, I really don't expect to be returning from Texas in this direction, but if that proves to be the case, and I need gas again, I'll certainly give your invitation a thought." The unfamiliar opacity of the code had flustered me.

The same was true with regard to that sign at the public end of the long private road leading back to the Mulberry. I turned from the windshield to the back of the Dodge van I was driving and quickly located my two terriers. Well, I *knew* where they were. I was not about to enter the property looking for them. Since doing so was the one activity which that curious sign emphatically forbade, I did not think myself prevented from entering the property for the purpose of simply seeing the grand old house, with its four corner bastions (each capped by small bell-shaped turrets), its hipped Dutch gambrel roof, and its English-bond brickwork. I drove straight past the sign, although its odd formulations continued to echo in my mind as I twisted in and out among the live oaks on my way to what proved an architectural epiphany. The house was a wonder, in perfectly restored condition, and I spent a good hour wandering around and around its exterior, photographing each façade and all of its many appealing details. I explained myself to a pair of workers, one busy about an old tractor along the lane and the other peaking her head out the back door. I was a teacher, I said, and wanted to photograph the house so I could use the slides in my classes. Was that all right? Both of the workers quietly nodded and let me alone as I went about my business.

An hour or so later, with two rolls of film exposed and a sense of justified risk and satisfaction floating in my mind like the evening mist gathering just then above the Cooper River, I turned the van around and started out the lovely road. It dipped down from the slight rise where the house stands, went past the man still at work on the old tractor, then curved off into the trees. Everything went very well until I saw, off in the distance through the live oaks, a small pickup. As it drew closer and closer, I became more and more tense. Would there be a confrontation? Would the truck swerve in front of the van and prevent my exit? Was I trespassing, in violation of the spirit of the sign if not its explicit text?

As the truck drew closer I could see that it contained a man, a young girl, and of course, a big dog. When the truck was perhaps thirty feet away, the man, smiling, flapped a lazy South Carolina wave as the dog jumped around in the cab and the girl quietly watched. I waved back and, with my own dogs still safely stowed in the rear of the van, went the length of the lane. I paused for a moment at that curious sign before turning onto the highway and heading to the next stop on my itinerary.

Then it hit me: the sign was a good old joke, most likely, a tongue-in-cheek put-down of all those many prying intruders who, when they got to the Mulberry, had been wont to claim that they were on the land "looking for their dogs." They may have had dogs, but they were not lost, or at any rate were not wandering the lands around the Mulberry. Although to those none-too-inventive sightseers the story probably seemed a capital dodge, it was no doubt a very different thing for the landowner, who had heard it over and over.

■ The point is, of course, that you will not fare well as a literalist in the world of signage. Most signs require a fitting sense of context and, even more, a tactful regard for the slippage inherent in all social situations and most texts. John Jakle and Keith Sculle have ample reserves of both traits. They will take a willing traveler precisely where they promise, with a running commentary that illuminates whole ranges of the ordinary landscape of America.

Preface and Acknowledgments

What follows is an exploration of how outdoor signage, as a means of communicating, affects the visualization of and, therefore, the use of landscape and place. By *landscape*, we mean the human habitat that surrounds us, the environment (both natural and human-built) that contains and sustains our every activity. By *place*, we mean those perceived centers of meaning that variously nest in landscape as settings or situations appropriate to, and thus supportive of, various kinds of social interaction. All human activity is place-oriented and place-contained. Accordingly, all social meaning carries implications of place-appropriateness. Thus, we read our surroundings for place meaning to properly "locate" ourselves in ongoing rounds of socialization. Signs, whether containing written words, pictures, or graphics, are deliberately displayed to be noticed and to be read. As such, they greatly facilitate the assigning of meaning to landscape and place.

Cultural geographers, among other students of landscape, have long talked of people learning to "read" landscape as a kind of scholarly, learned affectation. But the reading of landscapes is exactly what every one of us does all of the time. The surrounding environment stands like a text or a narrative to be constantly interpreted and reinterpreted. In particular, as we notice and attend to written sign messages, we quite literally do read the landscape, taking meaning from and assigning meaning to our surroundings in the process. Yet to date, few scholars have set about to explore just what it is that signs have traditionally contributed to the American scene. Certainly, signs orient, inform, persuade, and regulate, among other purposes. Clearly, modern society could not function without them.

Like everything else in our surroundings, signs stand waiting to be interpreted as "to whom it may concern" messages. In relation to most other symbol systems, whose meanings remain largely implicit, sign messages (what sign makers want viewers to know) tend to stand explicit. For example, the architectural styling of buildings tends to send messages more covert than overt. Spelling things out in words on signs, or using pictorial symbols, makes for less error of interpretation — or, at least, for narrower ranges of misinterpretation. In fact, most buildings require posted signs, both indoors and out, in order to function effectively. This is not to claim that specific signs always elicit like behaviors from every viewer. What signs encourage a person to do (or not do) varies from person to person according to his or her past experience, general competence at using signs, and, of course, personal intentionality (or behavioral predisposition) of the moment. Also, every sign is capable of communicating at several levels simultaneously. Beyond narrowly focused messages, they can communicate general, even vague, understandings. By their mere presence (or absence), they send messages about landscape and place.

Admittedly, the experiencing of landscape is multisensory. But landscapes, as built environments, are structured to be appraised initially by sight: things (or lack of things) in landscape, as well as their spatial organization, visually cuing place meaning. Kinds of people, their activities, the things they use in supporting what they do, and the timing or temporal pacing of their actions all serve to characterize place and bring meaning to landscape. Once we visualize landscape, we respond to it in various ways: for example, by deciding to enter and use specific places or kinds of place or, conversely, by deciding to avoid places. Even people who are not sighted operate in environments contrived to be seen. However, they must rely on hearing and other faculties to appraise settings. Only recently have public places in America come to sport signs intended to be read through touch. Important here is the notion that signs, as they are interpreted as visual images, form an important part of what might be called landscape visualization: the preliminary, if not continuing, appraisal of landscape for place meaning.

Place-making in America has been greatly influenced by changes in transportation. Through the twentieth century nothing more forcefully affected reorganization of the nation's geography than automobile use. Increased reliance on motor vehicles wrought decentralization of American cities, such that, fully automobile-dependent spread-out suburban landscapes dominated the nation by the twentieth century's end. Older pedestrian and mass transit landscapes were changed substantially as widened streets, new freeways, parking lots, and parking garages intruded to create suburban-like spaces even in big city downtowns. Thus, focus on the history of signs in America necessarily concerns the nation's emergent "automobility." Early in the twentieth century, outdoor signs were designed and positioned mainly to be read by people passing on foot. By the end of the century, most signs were intended to be read by motorists passing in cars.

We organize the book as follows. An introduction offers context for sign study as a form of landscape analysis. Substantive chapters are arrayed in four parts. In part one, chapter 1 and chapter 2 consider advertising, storefronts, window displays, and other signs as they aided business practice in pedestrian-oriented business districts early in the twentieth century. Chapter 3 concerns more the automobile and sign use, especially along the nation's newly developing highway-oriented commercial strips, where signs and architecture substantially conflated in signlike buildings.

Part two turns from commercial to other sign systems. Chapter 4 focuses on traffic control. Chapter 5 considers how signs can act to engender feelings of community belonging. Part three has but a single chapter. Chapter 6 explores how the reading of signs, and the displaying of signs, can contribute to a person's "sense of self." Part four considers American attitudes toward beauty and ugliness in landscape. Chapter 7 looks directly at what signs contribute to landscape by way of visual display. Emphasis follows the published literature's critical take on advertising signs as despoilers of the American scene. Chapter 8 deals with sign control and the extent to which public concern over sign aesthetics in the United States has translated into governmental regulation. Perhaps it is in the codifying of sign ordinances that signage, as an element of landscape and place, has come most forcefully to public consciousness. Finally, an epilogue draws together our various strands of inquiry.

We wish to thank the following people and organizations who helped us in our research endeavors: Susan Gas Barrow and the Ohio Department of Transportation in Columbus; Bill Barry and the Center for Maine History, Portland; Dianne Brooks and the National Safety Council Library; DeSota Brown and the Bernice Pauaki Bishop Museum, Honolulu; Paul Carnahan and the Vermont Historical Society, Montpelier; R. James Claus; Mary Kathleen Geary and the Transportation Library, Northwestern University Library; Stephen Gordon and the Ohio State Historical Society, Columbus; Bob Michaud and the Law and Reference Library, Augusta, Maine; Barbara J. Molansky and the Neon Museum, Las Vegas, Nevada; Kip Pope; Matthew W. Roth and the Automobile Club of Southern California; Mary Steiner and the Outdoor Circle, Honolulu; David Souder; Tod Swormstedt and the National Signs of the Times Museum, Cincinnati; and G. William Whitworth and the American License Plate Collectors Association.

Ergonomics Society, all rights reserved. The photograph from Phillip Tocker's article "Standardized Outdoor Advertising" appeared in *Outdoor Advertising: History and Regulation*, John W. Houck, ed. (Notre Dame: University of Notre Dame Press, 1969).

Special thanks to Jane Domier, who prepared the maps, and to Barbara Bonnell, who handled word processing. At the University of Iowa Press Charlotte M. Wright efficiently and politely managed the publication of this book and Clair James judiciously edited it. Both earn our praise.

Introduction

This is a book about signs and what various sign systems represent, both as material culture and as visual culture. By *material culture* we refer to the physical or structural aspects of signs. As a category of thing, signs have evolved over time to assume a wide variety of forms reflecting a wide variety of purposes: traffic regulation, geographical orientation, place labeling, and, of course, advertising in the promotion of products and services. By *visual culture* we refer to signs as something to be seen and thus as something to be read, both in and of themselves, as "texts," and as they contribute meaning to their surroundings. Signs represent an important communication medium through which people assert themselves in what might be termed symbolic interaction. Through written narrative and pictorial symbol, signs direct human intentionality, bringing degrees of social control to, and thus regularizing human behavior in, built environments of all kinds. They do so, in part, by enhancing place meaning. That is, they function as "to whom it may concern" messages designed to variously encourage, sustain, or change particular kinds of appreciations and behavioral predispositions.

This is a book about landscape, what we see when we look out over an area to survey our surroundings. We see landscapes as containing things, like trees, houses, roads, and signs, but usually in repeating patterns that speak variously of place: farms, highway roadsides, small town main streets, big city downtowns, and suburban subdivisions, for example. We learn and thus share culturally patterned ways of seeing landscape. We commonly differentiate rurality from urbanity, the planned and the orderly from the unplanned and the random. We think we know what is beautiful and what is not. But such conceptualization is constantly renegotiated as we interact with one another as social beings, both face to face in highly personalized ways and impersonally at a distance through one or another medium of communication as, for example, posted signs. People compete for social advantage by manipulating landscape as symbol. In consequence, landscape constantly changes.

Let us place ourselves in an imagined situation. Suppose we are driving out along a rural highway or toward a city's downtown. What would we see? What categories of things would we likely notice in the serial vision of our movement? It stands to reason that what the typical commuter sees in a city is not necessarily what captures the average tourist's attention. The purpose at hand and the extent to which movement is routine influence what is visually attended to. A tourist might savor the glimpse of a distant skyline toward which a freeway seemingly points, a particular skyscraper, perhaps, standing out as landmark. The commuter, on the other hand, might be focused more on the road and its traffic and thus remain oblivious to landmarks.

Comprehension of one's diverse visual surroundings necessarily involves categorization. Not only are specific objects named or labeled (identified through word association), but they are also sorted out and named according to the patterns that they create as ensembles. At the macroscale is landscape: the all-encompassing surround assessed in terms of broad patterns. Natural phenomena (sky, topography, vegetation) and cultural phenomena (buildings and roads) are seen to interrelate at a scale that might be termed "geographical." It is the scale at which geographers, as scholars, have traditionally focused. Places, as we will discuss shortly, nest within landscape as locales of special meaning. Buildings are, perhaps, the most readily identifiable kind of place. Consequently, most people conceptualize their surroundings, especially that which is human built, more at the "architectural" scale. It is the scale at which architects, as designers, and architectural historians, as scholars, have traditionally operated. Built environments are seen as constructed through the creation of individual buildings linked by and accessed from streets or roads. Some buildings vigorously call attention to themselves by size, ornamentation, or some other attribute and appear, accordingly, very much in the foreground of seeing. Other buildings, however, do not stand out and remain quietly as visual background.

Buildings, once noticed, break down visually according to architectural elements, especially those elements that carry functional implication. The articulation of openings across a building's facade is important, especially a building's main entrance, where "outside" and "inside" clearly interrelate. Buildings are nothing more than envelopes that variously enclose interior space to protect it and make it more useful.

Buildings are also noticed for their signage. Indeed, it may be the signs, rather than the architectural articulation, that stands out in many circumstances. At issue in this book is the posted sign, including, of course, signs quite unattached to buildings. In the American city, especially along commercial streets, signs forcefully announce building entrances, amplify facades, and sometimes articulate roof lines. But they also stand alone at the curbside to announce entrances to driveways and parking lots. Everywhere, whether in the city or in the countryside, signs orient and regulate movement. In our automobile-oriented world, traffic signs have become absolutely essential and are necessarily paid attention to, even if not obeyed.

In sum, signs are used to announce and give meaning to virtually every sort of place: signs on stores, billboards with their advertisements, stop signs at traffic intersections, address numbers on houses, signs identifying schools or churches, inscriptions on cemetery markers. What needs emphasizing, however, is that signs attract and hold attention not at the geographical scale of landscape, nor necessarily at the architectural scale of buildings, but at what we might term their own "signature scale." Like a person's signature on paper, signs endorse, verify, and confirm, and they do so very much at their own scale of recognition.

It is interesting to speculate whether people tend to move up or down our hypothesized scale of visual reference and whether the pattern reverses at night. Driving at night, the motorist necessarily attends to that which is lit, everything else being variously obscured in dark shadow. Traffic signs are necessarily illuminated or made to reflect brilliantly in auto headlights. With planned regularity, they loom and occlude in the serial vision of rapid nighttime movement in an effort to control speed, to cue turns, and otherwise to orient motorists to travel destinations. Important in cities, of course, is the illuminated street and also the lit building facade. In commercial zones, brilliantly lit signs attract attention in foregrounding stores and other places in the darkness. Beyond these lit objects and areas, however, the encompassing nighttime landscape, if noticed at all, tends to stand only as vague background. In the daytime, however, the reverse may hold true. The broad outlines of landscape are fully revealed. Architectural form is more completely evident. And signs, because they are no longer essential to the motorist's visual orientation, are less urgently attended to.

THE PLACE CONCEPT

Places are centers of behavioral expectation. They provide meaningful contexts for action in the ongoing streams of human activity constituting social life. A place is a finite setting that, because it contains a distinctive range of social activity, is seen to invite and thus sustain continued behaviors within that range. Place meaning, however, is something constantly open to negotiation. Thus, as a concept, place holds import not as a thing so much as a process. It is the process by which codes of conduct, place-sustained, are devised, contested, and changed over time as people define and engage new, but always place-oriented, needs. One way of linking the built environment to society's organization is to think in terms of place images. What places subjectively appear to be, and not necessarily what they are, predisposes individual behavior. We act on the basis of what we perceive or conceptualize reality to be. Of course, images of place must reflect the realities of the material world in some logical fashion, thus successfully underpinning behavioral expectation.

A place nests in landscape as a focus of interest, concern, or significance. It is assessed, consciously or unconsciously, as a basis for behavior. Place meanings are variously cued. Spatial or geographical context contributes; places are seen as having location. That is, they are meaningful in their proximity (or lack of proximity) to other places. They are seen as having areal extent and as having boundaries that demarcate, if not enclose. Buildings, as spatial enclosures, have solid walls that separate inside and out and through which human entrance and exit are highly focused (fig. 1). Over the long term, buildings stand not only as containers for human activity, but as memory prods that suggest behavioral appropriateness. When constructed, buildings are programmed to specific intended purposes, although those purposes and that programming invariably change with actual use as new needs and possibilities arise.

Places, therefore, are necessarily assessed in temporal as well as geographical context. Places age and may be seen as progressing through a life cycle from youth to old age. Places open and close in a sequencing of openings and closings set to diurnal and seasonal cycles. Places are seen as functioning for set durations of time. Places, especially those without clear structural containment, tend to be ephemeral. A chance sidewalk meeting between two people carries strong momentary place implication, meaning that quickly evaporates as conversation breaks apart. Individuals can "colonize" sidewalks or other public locations for personal use, creating a temporary sense of place in the process (fig. 2). Place meaning is also cued by the kinds of people who normally typify a place, and, as well, by their ongoing behaviors: not specific actions so much as usual or customary activities. So also do

FIGURE 1

Tavern in Challis, Idaho, 1978. Here is a place with clear architectural implication: facade, entrance, and projecting sign all communicate just how this place is to be entered. So also does the sign clearly suggest just what kind of place this is.

FIGURE 2
A ticket scalper near Riverfront Stadium in Cincinnati, 2000. Public sidewalks offer only temporary commercial opportunity. Place, as defined here, is ephemeral.

props or furnishings supportive of those activities, including signs, serve to cue place meaning.

Signs are often used as props. Transmission of place cues may be purposeful or accidental, complete or incomplete, accurate or inaccurate. Cues may be understood, missed, purposely ignored, or purposely misconstrued. Signs, however, usually function to clarify place meaning: to reduce whatever ambiguities other cues might sustain. They are part and parcel of place creation and place management whereby stewards of place seek to focus meaning in eliciting desired behavior from place users. Signs orient and identify; they can even variously set mood and elicit desire.

A place image comprises at least four components: belief, attitude, intentionality, and icon. By *belief* we mean the mental constructs that attach in suggesting that a place does, indeed, exist. By *attitude* we mean an affective and largely emotive orientation, a favorable or unfavorable stance toward an identified place. By *intentionality* we mean the way in which beliefs and attitudes fit into streams of ongoing behavior, intentionality implying behavioral predisposition. In peoples' minds, beliefs and attitudes attach themselves through intentionality to objects in landscape that might be called icons: actual things that symbolize. Icons cue complex thoughts about the kinds of satisfactions and dissatisfactions that one might expect in or from a place.

PLACES AND SYMBOLIC INTERACTION

Place meaning is implicated (sometimes overtly, but always covertly) in the communications that bind people together as a society. It is part of what sociologist George Mead termed "symbolic interaction."[1] A person, Mead recognized, has the ability to view his or her own behavior from the standpoint of others. Remembering past experiences, a person can anticipate the actions of others and can organize and direct behavior from the basis of felt preference. Life, therefore, is a constant negotiation whereby one interprets and is interpreted. People consider themselves assessed. More importantly, people assess themselves by fitting their behavior in with that of others, some sense of place-appropriateness always being operative in this assessment. When sense of place is strongly felt, people have only to reenact place-appropriate behaviors, the setting thereby seeming to invite or encourage certain behavior.[2] Once a situation has been interpreted, then like situations tend to be similarly interpreted; the world is thus construed and linked in terms of dependable place-types. As an aspect of personality, people develop styles of being in the world whereby preferences for particular kinds of places are cultivated.

Social situations might be thought of as temporal-territorial huddles where people communicate either face to face or indirectly through impersonal markers, or both. In this book, we emphasize the impersonal communication of signs. Signs are not directed at any one individual, but stand in public view for the benefit of anonymous others. Written language and pictorial depiction make sign communication possible. But verbal language is primary. Words are not mere symbols that happen along the way to understanding. They are the essence of the social process that understands. To fully comprehend something, people must verbalize it. Language is used when we communicate verbally with others, but also when we muse only to ourselves. To comprehend how places function in human life, scholars need to get behind the "rhetorical masks" that people use in place-assessing dialogue.[3]

Every experience is internalized in light of previous experience, and expectancies or anticipations are assigned to situations on the basis of such knowing. This process, called "construing," stereotypes things as "constructs." When a situation is new, it is construed using the experiences of previously encountered similar circumstances. Stereotypes are strengthened when expectancies are validated. When experience proves inadequate (when expectancies are not realized), then anxiety and even hostility can occur. More attention may be given to defining and verifying relevant constructs, or situations may be avoided simply because of the uneasiness generated by the less than fully known. Negative stereotypes also lead to situation avoidance. A person's belief system about a place (one's belief that it does or does not exist), a person's attitude toward a place (whether one views the place positively or negatively), and a person's intentionality, given the ongoing behavior at hand, derive from search for stereotype. Signs, for their part, serve to quickly prompt one or another form of stereotypic thinking. They are usually embedded in the built environment in order to enhance the familiar and thus reduce anxiety. In readily prompting the familiar, they serve to narrow the psychic distance between people and place.

Constructs may be thought of as object-defining, pattern-seeking, relationship-seeking, or system-seeking. However, these four orientations differ more in scale and in emphasis than in kind. Object-definition is essentially a classification process whereby a thing is labeled or assigned to a category. Words attach to thing. Additional understanding occurs when things are seen to form a pattern, to repeat themselves in recognizable ways. When relationships are observed and sets of relationships between patterns are seen to form a system, then even more understanding derives. When constructs nest one within another as, for example, when the stereotyping of people leads to the stereotyping of place (or vice versa), a construct may be object-defining for one purpose and pattern-defining for another. Similarly, a construct may be pattern-defining for people and relationship-defining for place, and so on.

Of central importance is the person's "self-concept": the various constructs that one uses to identify oneself as a kind of "object" relative to other people and to other things. A person's "self-concept," Mead wrote, consists of the "me" and the "I," the sense of "me" deriving from other people's reactions, and the sense of "I" representing one's response to those reactions.[4] The "I" interprets others, while the "me" is interpreted by others. Self-conceptualization is a vital element of place cognition. As people interpret both themselves and others, they give meaning to the places that they occupy, have occupied, or potentially might occupy. Persisting questions nag. We might phrase them as follows: Am I in an acceptable place? (Am I in the right place?) Do people here find me acceptable? (Do I fit in?) Beliefs and attitudes about place are rooted in such self-questioning, with behavioral predisposition, and sometimes behavior itself, following. Interactionists assume that human beings create the world of experience they live in. They do this by acting on things in terms of the meanings things have for them.[5]

Personal knowing and society as something knowable are mediated through widely shared meanings or sets of meaning. These meanings are defined, in part, by ideologies constantly reinforced through various means of communication, including use of signs. Ideology functions as an insinuating metanarrative. Messages come already interpreted in that they overflow with meaning. They have been reified as patterned regularities for interpretation and action. As stereotypes, they invite use with little questioning. They underpin taken-for-granted webs of significance that people come to rely upon in interacting with one another. Ideologies embed themselves in what passes as commonsense wisdom. For many, the term *ideology* carries derogatory implication, that of an insidious idea not totally obvious to the persons it influences.[6]

DEFINITIONS

In focusing on the material culture of landscape, cultural geographers and other scholars have treated a wide range of topics (houses, barns, field patterns, commercial building types, street patterns, etc.), but few have given serious consideration to signs.[7] In several previous books, we explored point-of-purchase and other kinds of signs at gas stations, motels, and fast food restaurants, places defined at the scale of the retail store fully oriented to automobile use.[8] But we kept signs, as a topic, subsidiary to architectural considerations, as have most authors concerned with landscape and place.

In recent years, the word *sign* has been co-opted by scholars focused on semiotics: the study of how symbols communicate as language.[9] Focused on the semantics of social "discourse," many scholars analyze written narratives (and, on occasion, even pictorial representations) for underlying social meanings. And yet signs (signs of actual material substance) provide one of the most important means by which social discourse is brought to and inserted into social life. Signs (actual signs) stand in landscape to influence thought, if not action. More so in some settings and less so in others, signs are what the managers of place rely on in asserting social control. Through signs, social agendas can be imposed and maintained. And, of course, through signs, social agendas can also be challenged.

Perhaps, when Americans think of signs, they think primarily of advertising. American society is, after all, organized around an essentially capitalistic economic system with an inherent need to sustain production through sustained consumption. The business of America is business, pundits like to say, and advertising signs, for their part, serve the nation as business catalysts. The annual amount spent on outdoor advertising (specifically billboards and mass transit posters) topped $5 billion in the year 2000.[10] In 1997, there were nearly a half-million billboards lining the federal-aid highway system alone.[11] The placing and tending of billboards was very big business. Infinity Outdoor, then the nation's largest billboard operator, controlled nearly a million signs, and the nation's second largest, Eller Media, some 750,000.[12] A simple fact of social empowerment is here evident. The power to persuade in America, through outdoor signage at least, is substantially concentrated in a relatively few corporate offices.

What is a sign anyway? Perhaps we should stop and offer more formal definition before continuing to consider what it is that signs do and how it is that they do it. Let us turn to *Webster's New International Dictionary*. As a noun, the word *sign*, according to *Webster's*, carries two closely related and fully complementary meanings: that of "symbol" and that of "indication." A sign, as students of semiotics emphasize, is "a conventional symbol or emblem which represents an idea." Closer to our use of the term, however, is "a lettered board or other conspicuous notice, placed on or before a building, room, shop, or office to advertise the business there transacted, or the name of the person or firm conducting it."[13] It is interesting that the commercial or business roots of the term are emphasized. *Webster's* further defines the word *signboard* as "a board for or bearing a notice or sign, originally of a shop or inn."[14] But in the broader sense, *Webster's* also defines a "sign" simply as "something serving to indicate the existence of a thing." Accordingly, signs can carry temporal significance as "presage" or "portent" relative to the future, or as "trace" or "vestige" of the past. As a verb, "to sign" can mean the following: to "place a sign upon" (as in consecrating), to "symbolize" or "betoken" (as in marking), to "subscribe a signature" (as in signing one's name), to communicate by signs (as, for example, deaf people using "sign language"). Missing, however, is any notion that environment can be "signed," as with the placing of signs for social purposes.

Turning to the sign industry for definition, we are helped by James Claus and Karen Claus, a geographer and a legal scholar, respectively, who have devoted their careers to researching signs for various industry research groups. They define signs as "any structure, device, [or] . . . visual representation intended to advertise, identify, or communicate information to attract the attention of the public for any purpose." Thus included are "symbols, letters, figures, illustration, or forms painted or otherwise affixed to a building or structure and any beacon or searchlight intended to attract the attention of the public for any purpose and also any structure or device the prime purpose of which is to border, illuminate, animate, or project a visual representation."[15] Outdoor advertising (their specialty) is very much of and in the public arena. Full comprehension of how signs work, they argue, necessitates concern with the contexts of landscape and place.[16]

We would emphasize what it is that signs do. Whether for commercial purposes or not, signs perform the following functions, sometimes singly and sometimes in combination. With regard to built environment, they (1) identify, (2) persuade, (3) orient, and (4) regulate.

First, some signs function primarily to identify places. That is, they make places more legible or visible, for example by cuing entry points, defining perimeters, or otherwise marking some locus of place attention. They serve, in other words, to identify a place as a meaningful locale. Sign messages, communicated verbally or graphically, denote. But signs, by their size, color, and level of illumination, among other design attributes, can also connote. Pictured is a street intersection near the center of a small Indiana city (fig. 3). A gasoline station sign identifies a particular place where certain goods and services are available for purchase. But the logo also announces that the station is affiliated with the product line of a specific corporation and thus is part of a network (or chain) of similar company-operated establishments. This implies that the discriminating customer will find at this establishment expected brand-specific products and an expected level of service satisfaction.

Second, some signs function more to persuade. They stand not just to inform, but to excite, entice, and, ultimately, convince. Their purpose is to engender, through some implied line of argumentation, entreaty to action (or inaction), either specific to a given place or to places generally. In the photo, the Standard Oil logo that marks the gasoline station's driveway entrance carries beneath it price quotations for various grades of gasoline, price being an important dimension of place meaning by which gas stations are patronized.

Third, some signs function more to orient people toward places. That is, they offer directional advice as to where things are located and how one might proceed to reach those locations. In other words, they steer. As pictured in Figure 3, directional arrows, paired with route numbers, cue motorist turning decisions. Another sign points to the local hospital.

Fourth, some signs serve primarily to regulate behavior. They regulate the type, intensity, or duration of a desired action. Pictured are traffic signal boxes that, in their timing, order vehicular and pedestrian

FIGURE 3
Street intersection in Frankfort, Indiana, 1997. Signs function to variously identify, persuade, orient, and regulate vis-à-vis place meaning in landscape.

movements through the intersection. In a broader sense, however, these stop-and-go lights, along with the route markers, imply an important kind of territorial jurisdiction. It is not the influence of a private corporation that is symbolized, but, rather, the ability of state and municipal governments to define and regulate public space.

■ Signs not only have physical substance, but are configured and located specifically to be noticed, thus contributing to the structuring and functioning of landscape as place. They carry symbolic value, contributing to the sustained symbolic interaction that sums as American society. With signs, Americans sustain group- and self-identities, regulate behavior both in public and private spaces, sustain consumption (and thus production) in an economic sense, and realize a host of other social purposes. In attracting attention, signs have an impact on how landscapes are conceptualized, especially regarding place identification, place orientation, place assessment, and, ultimately, place use. Sense of place is an essential of social being since all human behavior does, in fact, "take place." In the chapters that follow, we explore various ways that signs contribute to defining place within landscape. We focus especially on America's growing dependence on auto use and its impact on built environment.

Signs have evolved as an important form of material culture. As such, they are worth considering as cultural artifacts quite apart from their social implications. In the United States, signs are more frequently encountered, perhaps, than any other sort of landscape feature, and yet they have been little studied by students of material culture. They stand alone or are attached to, and thus are used to complement, other well-studied features of the built environment: stores, factories, barns, fences, and even houses. Signs do have social meaning as they function as a form of visual culture. In their being seen, they are intended variously to cue thought and action. They are intended to influence human behavior either in a general sense, as applicable anywhere or, in some more specific sense, as pertaining to a certain place or kind of place. Signs stand quite literally to be read for their meaning. As such, they act to promote particular social agendas. In that promotion, they become instruments of power in advancing some social interests over others.

PART ONE COMMERCIAL SIGNS

1 : Signs Downtown

The signs of commerce brought much visual excitement to American cities, especially in central business districts. What spoke forcefully in America of urban vitality was not architecture alone, nor architectural ensembles concentrated along streets at the scale of landscape, but buildings, plain and simple, as they were variously signed as distinctive places for business. America's cities were largely commercial affairs. Few were laid out with picturesque tree-lined boulevards anchored by monumental buildings. They were not structured, in other words, to reflect directly elite social and political power, as was more the case in Europe. Rather, American cities were laid out on utilitarian street grids essentially as real estate speculations.

Throughout much of the nineteenth century, Americans knew their cities primarily as pedestrians. Thus, signs were configured primarily to communicate to people walking on sidewalks. With the increased importance of mass transit and certainly with the twentieth-century arrival of the automobile, commercial signs necessarily had to communicate over increased distance: building facades were seen more and more through transit car windows and automobile windshields.

To walk is to be closely involved with one's surroundings through sight, sound, smell, and touch. However, when cocooned in an automobile, close encounter with city space is much reduced. The detail of slow multisensory intake gives way substantially to that visually experienced through the eye's scanning: the city reduced in its conceptualization to rapid gaze. In the automobile city, signs are absolutely essential to the "reading" of landscape for place meaning.[1]

STOREFRONT SIGNS

At the time of the American Revolution, commercial and residential functions were very much mixed in cities like Philadelphia and New York. Most artisans and merchants worked where they lived, usually in a row house closely sandwiched between other like structures along a street. Shops, where customers entered directly from street or sidewalk, were necessarily located at a building's front, family living space being pushed to back rooms and upper floors. A signboard protruding above a door might announce the proprietor's name or suggest, through a painted or carved likeness, the product or service available within. Goods might also be displayed in a front window — door, sign, and window combining as invitation to enter.

By the Civil War, business and residential activities in cities tended more to be separated building to building, although different kinds of buildings remained substantially mixed within neighborhoods. Only in elite residential areas were business buildings totally resisted, and only in the new central business districts, where commercial activity concentrated at a vast new scale, were houses largely missing. Banks, department stores, and office buildings (which housed insurance companies, newspapers, and trade associations, among other enterprises) represented distinctly new building types. But the row house precedent still influenced most retailing, which remained primarily the bailiwick of the small entrepreneur. Shops still elbowed one another closely, but often with fascia signs and display windows of plate glass stretched across entire storefronts. By World War I, the scale of downtown retailing across America had substantially increased, small shops giving way to large specialty stores. Skyscrapers soared to great heights, offering opportunity for giantism in rooftop and other sign displays.

Taken as a whole, the signage that Americans actually saw and responded to in big city downtowns early in the twentieth century was quite complex, especially along major retail streets. Old signs often remained as new signs were introduced, creating a palimpsest of sign art, often with substantial historical depth. In streets given more to wholesaling or warehousing, sign innovation was usually less evident, a function of more conservative business practice given that wholesale firms did not compete with one another in quite literally attracting customers "off the street."

Sign conservatism also played among more prestigious businesses right at a city's heart. Banks, insurance offices, and other businesses that could afford to do so invested in expensive, highly durable signs configured to appear, and to be, long-lasting. It was not inertia at play, but a quest for symbolism suggestive of business strength and financial security. Carved in stone or engraved in brass, molded in bronze or etched in polished glass to be bolted onto building fronts, such signs identified businesses as part of a city's commercial elite. They spoke in understated terms about business excellence. They were made deliberately tasteful.

It was the electric sign, however, that captured maximum attention in commercial districts as the twentieth century progressed. Electric signs not only communicated forcefully in the dark of night, but were greatly enhanced over other signs when illuminated in the day. They spoke of businesses being up-to-date and changeful in modern times. For retailers, dependent on walk-in custom, electric signs became absolutely essential. Observed journalist George Williams, writing in 1909 in the new trade journal *Signs of the Times*, "The electric sign is the one object of attraction that people cannot and do not want to avoid. It is a democratic institution. It is admired by the saint and the sinner, the aristocrat and the plebian, the cultured and the vulgar, the scholar and the illiterate."[2] Electric light had universal appeal. But, more to the point, wrote Bury Dasent in the *Journal of Electricity, Power, and Gas*, "The fiery electric signal has become in truth 'the sign' of the progressive merchant, and takes its place as a matter of course in the equipment of every high class and up-to-date establishment."[3]

Because of the signs' brightness, motorists could more readily see electric signs in the day. This was especially so when signs were projected out over sidewalks perpendicular to storefronts. From a car motoring down a street, building facades were viewed at an oblique angle, making fascia signs, especially small ones, difficult to decipher at a glance. When electric signs were additionally set to blinking on and off, they attracted even more attention. Called "flashers," they seemed to compel notice and were actually cheaper to operate as they consumed less power.[4]

Nonetheless, electric signs tended to be fragile and required frequent maintenance, if only to replace burned out lamps or tubing. Technical innovation was constant, thus inviting sign replacement through upgrading. At first, most storefront electric signs were lit by exposed incandescent bulbs arranged to spell out words. Use of translucent materials lit from behind came to the fore through the 1920s. Neon signs with exposed luminous tubing obtained widespread popularity in the 1930s. But most popular of all storefront signs, undoubtedly for their lower costs of construction and operation, were metal signs of highly reflective porcelain-enameled sheeting that were merely floodlit at night (fig. 4). Whatever the type, sign makers found themselves concerned with the following attributes of sign design: overall form (including size and shape), pattern (including lettering and pictorial representation), color (both hue and tone), and motion (involving flashers or other mechanisms).[5] In that regard, fads and fashions in sign design came and went over the decades in a decided drift toward lower costs and enhanced sign legibility for motorists.[6]

OUTDOOR ADVERTISING

Off-premises signs that advertised goods and services, usually by brand name, also came to line downtown streets. After 1900, posted bills (or posters), photomechanically reproduced with powered printing presses, led the way, as did large signs painted on the walls of buildings. Billboard advertising involved multiple printed sheets pasted together to create large integrated images.[7] Such signs were directed toward a fully anonymous "public": a hypothesized market distanced in time and space from an actual business transaction. Their purpose was not to excite immediate purchase, but to instill awareness (if not create a sense of impelling need), predisposing customers toward eventual consumption.[8] The increasingly affluent American masses, with their increased leisure time, were, of course, the consuming public that the advertising industry sought to manipulate.[9]

Capacity to buy widely available goods and services, however much it vindicated capitalism in the debate over public responsibility versus private enterprise in early twentieth-century America, met with reticence in some established social circles, especially when it came to bill

FIGURE 4

Floodlit porcelain enamel signs in Covington, Kentucky, 2000. Highly durable, requiring less in maintenance than most other kinds of storefront sign and yet also highly legible in the night, porcelain enamel signs continued to be popular after World War II.

posting. Critics saw billboards as blighting public space for private gain. Disgust with billboards was also rooted in the growing tension between social classes, especially in regard to newcomers recently arrived from abroad. Immigrants, unable to read English-language newspapers and magazines (and the wordy ads that they contained), could be readily persuaded by billboards, which relied more on pictures, and on short, easily-interpreted phrases.[10] Through billboard pictures variously positioned in urban space, the illiterate probably were made to seem more a part of America's social mainstream, much to the consternation of social purists.

J. Walter Thompson, who pioneered advertising in elite magazines beginning in the 1870s (and whose agency consistently earned the highest billings in the industry for fifty years beginning in the 1920s), added billboard advertising despite lack of interest from many advertisers. For example, Eastman Kodak pitched its low-priced and easily operated cameras mainly in genteel magazines and reported as late as the early 1920s that it seldom employed billboards and was categorically opposed to streetcar advertising, the two bill-posting venues believed most persuasive with the working classes. In contrast, the manufacturer of Sapolio, a popular scouring power, took very early to advertising cards, posting ads initially on New York City streetcars in 1884. A subsidiary firm became one of the first advertisers to standardize postings as to size. Around 1900, advertising agents, who had moved beyond brokering space for advertisements in the print media to include copy writing and artwork design as well, also began to handle outdoor advertising.[11]

In the throes of the nation's growing affluence and cornucopia of goods and services, sign makers began implementing the rudiments of what communication theorists have recently come to call the semiological triangle: communication conceptualized in terms of a signifier, a "sign" (or referent), and a signified.[12] People, as signifiers, communicate experience to one another using shared language, for example, spoken words as "signs," "signs" that stand in as referents for that which is being referenced or signified. Thus, signifier, signified, and "sign" are all interlocked, and leaving one out makes full comprehension of how people communicate, including advertising's exhortations, impossible. Functional differences between the products of different firms (potentially reduced to quantitative and testable distinctions such as cost, dependability, and efficiency) were inserted as signifiers into common language through advertising. Persuasions, carried over from the time when producers and consumers negotiated with one another directly, were recast for the impersonal mass market.[13] For example, ads promoted products by associating them with certain kinds of people depicted in certain kinds of places, for example, fashionably attired socialites in country club settings.[14] Rather than promoting superior product quality or dependability, advertisers increasingly emphasized a product's life-style implications, promising personal therapy if not improved social status.

Images eventually outstripped words in all advertising copy, even in magazines. Pictures came to dominate. Such images, and the associations they engendered, reverberated in layers of meaning: the signified of one stage of comprehension becoming the "sign" for another, on and on. To borrow from sociologist William Leiss and his associates, advertising is not just a business expenditure. Its creations "appropriate and transform" a vast range of social ideas. Advertising unifies as discourse through and about objects, bonding them together as persuasive images of well-being.[15] A geographical corollary complements the sociological claim, since advertisements situate people in places, both metaphorical and literal places.[16]

By the beginning of the twentieth century, outdoor advertising signs had grown very specialized. Five sign categories were identifiable according to size, composition, and location — to use the industry's own taxonomy of the time. Posters were most common. Poster subtypes included lithographed paper sheets or panels, streetcar cards, cards used in display windows, and cards tacked onto convenient surfaces like posts and walls. Hand-painted signs, another category, was divided into the following subtypes: bulletins — with subcategories (city, highway, and railroad locations, for example), painted walls, signs over business entrances, and small signs. Three other types included electric signs (both storefront signs and large "spectaculars"), window dressing signs, and miscellaneous signs. The latter included, for example, suspended banners, flags, sandwich boards, and truck graphics.[17] Of the various sign types, those that attracted maximum attention in big city downtowns were undoubtedly the large wall signs, the billboards, and the so-called electric sign spectaculars.

PAINTED WALL SIGNS

By 1900, most American big city central business districts sported large signs painted on the brick walls of businesses and other buildings. Required was a sidewall unbroken by windows or other openings that, for lack of an adjacent obstructing structure, was visible from a public way (fig. 5). With such signs, landlords could extract additional rent from their properties and merchant tenants could reinforce sales. But the most benefit accrued potentially to the sign company as creator of the display, and, of course, to the advertiser whose product-line the sign promoted.

To be glimpsed in passing, painted wall signs usually offered only very short, even cryptic, written messages, slogans predominating. Most were highly pictorial: a picture of a product (or, more likely, its packaging) or a picture of the product being consumed or used. So also were advertiser logos prominently featured by way of reinforcing brand identity. The main purpose of such outdoor advertising was to excite a sense of need. It was the "quick take" that counted. Large signs were difficult to ignore in the context of a city street. Such outdoor advertising, wrote one journalist, was "a flash of color, a pleasing impression, a staccato message, again and again for day after day."[18] Customer predispositions were expected to well up through repeated exposure.

BILLBOARDS

The term *billboard* appropriately designated signboards upon which paper poster panels were pasted either singly or in combination. The advertising industry sought to emphasize the positive by characterizing billboards as art and bill posters as artisans. Thus, billboards were said to visually enhance the urban scene, turning the downtown street, it was frequently argued, into something like an outdoor art gallery. But more often billboards were seen as obstructing scenery rather than as

FIGURE 5
Ideal location for a painted sign at street level. Source: Wilmot Lippincott, Outdoor Advertising, *p. 188.*

contributing to it. When clustered in large numbers, they were criticized for bringing visual chaos to downtown scenes. Almost from the beginning critics called for their removal. Resistance to their indiscriminate use divided the outdoor advertising industry itself, one faction committing to self-regulation and another holding out for complete laissez-faire.

In 1900 the Associated Bill Posters' Association standardized posting practices. Sheets, standardized at forty-two inches by twenty-eight inches, were to be mounted on boards capable of holding three, eight, or sixteen sheets. However, the preferred size came to be the twenty-four-sheet poster panel (four sheets high and six sheets long) adopted by the association in 1912, a size that resonated better with motorists. Standardization, at whatever size, suggested industry restraint. As large manufacturers with big advertising budgets were among those approving the decision, standardization came to be accepted even by the most recalcitrant of poster firms. Also, entirely wooden sign structures (supports, frame, and poster boards included) began to be replaced by steel structures. Proponents argued for their increased longevity (three to four times that of wooden signs) and, more importantly, their modern appearance.[19]

Outdoor advertising firms, of course, located billboards throughout cities, not just in downtowns. Yet through 1930 most firms did concentrate the bulk of their boards either in central business districts or at their immediate approaches along major thoroughfares. When the automobile came to the fore, as we emphasize in chapter 3, more and more billboards appeared along outlying thoroughfares, and along rural highways as well. Rarely were billboards rented singly. Rather, their use was usually part of a marketing package, called a "showing," whereby a specific poster display, or set of displays in rotation, was shown systematically across the boards of a given locality for a stipulated period of time. National advertising campaigns encouraged the more aggressive outdoor advertising firms to create branch offices in order to service directly leading markets. Billboard rates were calculated on the basis of estimated daily effective circulation, the number of vehicles and pedestrians that potentially passed an array of signs on a daily basis. Firms networked with one another in showing given bulletins across their respective marketing regions.

Thus, a manufacturer, like the Eureka Vacuum Cleaner Company in 1924, could launch a marketing campaign using billboards in support of its nearly four thousand dealers nationwide, the same poster appearing in some 150 different urban areas (fig. 6). Every housewife was

FIGURE 6
Billboard display used by the Eureka Vacuum Company in a nationwide marketing campaign in 1924. Source: Poster *15 (May 1924), p. 8.*

offered five days' free use of a Eureka vacuum cleaner for spring cleaning. The billboard campaign was carefully coordinated with ads in local newspapers and in such magazines as the *Saturday Evening Post*, *Ladies Home Journal*, and *Good Housekeeping*. The firm sold 228,000 machines in 1923, or one of every three vacuum cleaners sold in the United States. That figure represented a rise from one in four in 1922, one in six in 1921, and one in ten in 1920. The billboards were given most of the credit for the increase.[20]

THE OUTDOOR ADVERTISING INDUSTRY

Advertising posters were first made available through numerous one- and two-person shops throughout the United States. In 1865, 275 bill-posting companies, employing from two to twenty employees, operated across the nation. Circuses (for example, that of P. T. Barnum) and local theaters or "opera houses" sustained demand for small posters at least seasonally. But the advertising of manufactured products, with their much larger earning potentials, spawned truly viable year-round trade. The growth of large firms occurred first in the nation's larger cities, where financial rewards were greatest.

Boards leased for bill posting were first popularized after the Civil War, the term *board* (as in *billboard*) reflecting the fact that most were wooden structures literally constructed out of boards. In 1867 the

Bradbury and Houghteling Company was launched in New York City, ultimately to become the first nationwide sign-painting service. The firm's founding coincided with the opening of the city's first elevated railroad, the right of way of which provided opportunity for posted as well as painted advertisements. In 1872 Kissam and Allen, also of New York City, became the first company to build, own, and rent poster panels.[21]

The International Bill Posters Association of North America formed in St. Louis in 1872. Although it ceased operations only twelve years later, the impulse for control from within the young industry was made clear. In 1875 the first state organization was established in Michigan, with trade groups in Indiana, Minnesota, New York, Ohio, and Wisconsin operational by 1891. In that year, representatives of a number of bill posting companies, along with representatives from several of the state organizations, met in St. Louis to form the Associated Billposters' of the United States and Canada, the idea being to better coordinate the business activities of members. Territorial rivalries, which in New York City had nearly erupted in physical violence, and the question of where and where not to affix posters (fences and walls versus sign structures, for example) were among the first issues debated.[22]

Various other organizations affected the outdoor advertising industry. Soon after the Associated Billposters' group was formed, its directors encouraged William H. Donaldson, son of an early poster printer in Cincinnati, to transform his magazine, *Billboard*, into that organization's reporting vehicle. Instead, Donaldson made the magazine into an industry-wide journal, a role that it continued to play until ceasing operation in 1931. Stemming from a disagreement at a meeting in Detroit in 1906, twenty members of the Associated Billposters' Association launched a breakaway organization, the Advertising Painter's League of America, to mollify a public backlash against the industry's often offensive aesthetics. Donaldson founded *Signs of the Times* for the breakaway organization; it became, and remains, the sign industry's principal trade journal. The National Outdoor Advertising Bureau was organized in 1918 to service potential advertisers, more specifically to encourage their use of outdoor advertising signs, especially billboards. Many general advertising agencies still did not fully appreciate how the outdoor medium could be used, a circumstance that the bureau was specifically designed to counter.[23]

A few big sign companies came to dominate poster printing and the sale of poster space, a situation typical, perhaps, of a matured industry. Brisk rivalries were sustained, however, when federal antitrust legislation and several federal district court decrees enjoined companies, such as the mammoth General Outdoor Advertising Company, from conspiring to limit competition. R. J. Gunning, who had entered outdoor advertising in Chicago in 1873, was among the first of the industry's powerful moguls. By 1906 his billboard construction and rental firm had plants in Cincinnati, Louisville, St. Louis, Kansas City, Omaha, and Minneapolis–St. Paul, among other cities. Gunning acquired considerable power within the industry when he sold his printing operations to focus on the selling of advertising not only through rental billboards, but also in newspapers and magazines, linking, thereby, outdoor ads directly with print ads. In 1875 Thomas Cusack founded another Chicago firm. When Cusack's firm merged in 1925 with the thirty-year-old Fulton Group, it created the largest outdoor advertising firm up to that time. The General Outdoor Advertising Company came into being when this firm was combined, in turn, with the O. J. Gude Company of Brooklyn.[24] Foremost on the West Coast was the Foster and Kleiser Company, which began in Portland and Seattle in 1901. In 1925 the Outdoor Advertising Association of America was formed to service the entire industry, overriding, in the process, sign painter tendencies not to cooperate with one another in the belief that artistic skill, rather than marketing acumen, spoke above all else.[25]

ELECTRIC SIGN SPECTACULARS

Billboards could be floodlit to communicate in the dark of night. Required were bright lamps (which General Electric's new Mazda incandescent bulbs provided), highly reflective sign surfaces (which favored painted metal over poster paper), and reflectors (which eliminated the glare of exposed lamps). But, despite the bill posters' claim to the contrary, floodlit bulletins simply were not as bright nor as eye catching as large electric signs. "By 'electrical advertising,'" wrote G. H. S. Young of the Edison Electric Company in 1912,

> I refer to electrically operated and illuminated signs . . . which are in most cases placed on roofs of buildings on busy streets, and which, by reason of the letters and effects being carried out by many small electric lamps which "sparkle" and flash their message against a dark night sky, "compel" the attention of all passers by.[26]

Sky signs were supported on heavy steel frameworks, scaffolding which was usually left fully open in the early years (fig. 7). Sign installers, who were originally dubbed "sign hangers," quickly acquired

FIGURE 7
Sky sign at New York City's Times Square, 1925. Source: "Circulation Thirteen Cents per Thousand at the Crossroads of the World," Signs of the Times *50 (June 1925), p. 28.*

the building skills of engineers, their fabrications being building-like in scale and in complexity of construction.

It was in and around New York City's Times Square that the nation's largest concentration of sign spectaculars evolved. Among the more memorable sign displays before World War I was Clicquot Club's chariot race placed high above the Hotel Normandy at Broadway and Thirty-eighth Street. The race and its effects played out in exposed incandescent electric lamps that flashed in complicated sequences. "The leading horse was hard pressed by others," wrote S. N. Holiday in *Signs of the Times*. "Dust clouds, flying colors, racing horses, dangerous corners all depicted a thrilling arena scene."[27] At Times Square, one could have seen in 1917, as Arthur Williams put it, "a veritable electric moving picture gallery."

> There is the Djer-Kiss sign, with its exquisitely colored flashing bouquet of flowers; the Warner's Rust Proof Corset sign, where in electric light a man is shown sprinkling water on a corset of that make; Corticelli Silk, with the now famous kitten entangled in its meshes; the bird flapping its wings in the Anheuser-Busch Budweiser sign; the Pebeco Tooth Paste sign, with its moving border of

advertising slogans; the big golden sign of Chalmers Underwear; Coronet Dry Gin, pouring its electric streams into two glasses . . . and many others.[28]

Quoted statistics always emphasized the size of Times Square signs, and rightfully so. Sign spectaculars were erected along the principal downtown streets of most American cities, but those at Times Square were invariably larger, besides being concentrated in greater numbers. Numerous corporate marketing campaigns used a Times Square display to anchor advertising nationwide, pictures of a Times Square sign being used extensively in newspaper and magazine ads. The Wrigley Company's first sign spectacular at Broadway and Forty-fourth Street in 1917 was fifty feet high and two hundred feet wide and required some seventeen thousand incandescent electric lamps. Six eighteen-foot, acrobatic "spearmen" (for spearmint gum) went through a series of setting-up exercises. The company's 1936 sign, which portrayed a mermaid with fish swimming under water, was 75 feet tall and 188 feet wide and used 29,500 incandescent bulbs as well as 1,084 feet of neon tubing. In addition, its construction required 86,661 pounds of structural steel, 275,150 feet of wire, 4,140 feet of conduit, 500 feet of high-tension cable, 597 pounds of solder, 725 pounds of electric cement, 470 pounds of insulation tape, 1,298 fuses, 28 transformers, and 120 gallons of paint, among other materials (fig. 8).[29]

Sign spectaculars functioned to advertise the corporate products of the new consumer revolution, but, when concentrated in large numbers as at Times Square, they could also make important statements about community and place. Times Square was the busiest pedestrian area in New York City, several subway lines intersecting there just blocks west of Grand Central Station. But Times Square was also an important tourist attraction, no visit to New York City being complete without witnessing its nighttime sign extravaganza.[30]

Across America, giant signs could also be found in limited numbers along major retail streets, along city waterfronts (especially along railroad approaches), in wholesale and warehouse districts, and even in industrial areas. In Louisville, space for large electric signs was created on a specially erected framework visible to railroad passengers crossing the Ohio River by bridge, the words "Gateway to the South" blinking across its top. Manufacturers frequently topped factory buildings with giant signs. In 1923 Colgate erected a sign atop its Jersey City plant across from Lower Manhattan. With the word FAB spelled out in letters

FIGURE 8
Wrigley's Spearmint gum sign spectacular at Times Square in New York City, 1936.

fifty-six feet tall and twelve feet wide, the sign could be read upward of eight miles away and seen, as a brightness in the night, upward of thirty miles.

■ From storefront signs that alerted pedestrians up close to business opportunities, to giant wall signs, billboards, or sign spectaculars that spoke as well to transit riders and to motorists, signs in big city downtowns underwent constant change. New kinds of electric signs, for example, evolved as new lamp technologies were introduced: exposed incandescent electric lamps giving way to the brightness of exposed neon tubing and, then, to cheaper-to-operate fluorescent tubes boxed behind backlit panels. But of all the new technologies to affect signage in big city downtowns, it was the coming of the automobile, pure and simple, that had the biggest impact. With the increased importance of motoring, storefront signs became larger, were electrically illuminated, and were increasingly hung projected out over sidewalks in order to better catch the motorist's rapidly moving eye. Nonetheless, it was the large electric signs that captured the public imagination, especially as they promoted branded merchandise, itself symbolic of the nation's emergent modernism.

Few could deny that signs added visual excitement to central business districts. Of course, as critics claimed, too often this visual excitement fringed on visual chaos. The question was whether signs went too far in attracting and exciting the eye. Even in excess, however, signs humanized city streets with their exhortative messages. Although they spoke impersonally to the passerby, the messages conveyed could excite, arouse, reassure, convince, and, in various ways, make city space more readily comprehensible. As signs catered to consumer impulses, they informed personal identity. As they inspired admiration, they could even speak of civic pride and community belonging.

2 : Signs on Main Street

Ever since Sinclair Lewis's classic novel *Main Street* appeared in 1920, small towns have been stereotyped for their parochialism. The novel is the story of Carol Kennicott, who arrives in a small Minnesota farm town from the nearby Twin Cities as bride of the town doctor. Character and plot development in the novel hinge around Carol's various attempts to improve her adopted hometown by making it more urbane. But ultimately the town, Gopher Prairie, brings her down to its level, its parochialism sustained. On her arrival, Carol sets out to explore: a walk that eventually brings her over to the town's Main Street with its concentration of stores. Lewis wrote:

> From a second story window the sign "W. P. Kennicott, Phys. & Surgeon" gilt on black sand. A small wooden motion-picture theater called "The Rosebud Movie Palace." Lithographs announcing a film called "Fatty in Love." Howlands & Gould's Grocery. In the display window, black overripe bananas and lettuce on which a cat was sleeping. . . . Flat against the wall of the second story the signs of lodges: the Knights of Pythias, the Maccabees, the Woodmen, the Masons. . . . A jewelry shop with tinny-looking wrist watches for women. In front of it, at the curb, a huge wooden clock which did not go.[1]

Thus, signs not only identified stores and other locations, but, in their totality, were seen to characterize the place.

In reality, small towns, especially those of the Midwest, where farming remained relatively viable, were not totally parochial. They were, in fact, highly imitative of the nation's big cities, at least in the configuring of their principal shopping streets. Satire was generated by critics not so much around the lack of urban amenities, but by the scale of their small-town introduction and the sometimes naive local enthusiasms which accompanied. Electric signs were but the first step in small town improvement, argued one journalist in 1913. "Pretty soon the town is ablaze," he wrote, "and it is noticed that the merchants develop civic pride enthusiasm, which is passed through the newspapers to the public. The show windows begin to develop a stronger attraction than the fireside, and the municipality graduates from the town class into the city class." In conclusion, he noted: "The electric sign has awakened more slow towns and brought to life more dead streets than any other force in municipal affairs."[2]

Suppose you lived in a small town, an editorial in the *Signs of Times* intoned. "As you come downtown after supper on Saturday night, and everybody is out, how cheerful it looks down toward the square. The show windows are bright and attractive; the electric signs contribute their part. And how the people seem to enjoy it. The brightness and good cheer stimulate the desire to spend money."[3] Spending money was the new patriotic thing to do after World War I. Flowering in the 1920s was an American consumer society quite unlike that ever before known. Discretionary spending (beyond the minimal necessities of food, clothing, housing, and health care) increased from 20 percent to 35 percent of America's purchasing power during the first three decades of the twentieth century.[4] Implicit prosperity meant a shift from purely utilitarian goods to symbolic goods. Through their packaging, display, and advertising, observed historian Gary Cross, consumer goods came to embody a distinct and eventually dominant alternative to distinctly political, and even religious, visions of what America should be.[5]

The new commodity that was the bellwether of this trend was the automobile. In America, the motorcar did not long remain an exclusive plaything for the well-to-do. Mass-produced by Ford, General Motors, and other companies, there were some 8 million motorcars registered in the United States in 1921. Ten years later motor vehicle registrations totaled over 26 million, 23 million being private cars.[6] Besides new signs, another way of symbolizing and, indeed, of promoting the new America on Main Street was to improve the street itself, thus making it more accommodating to shoppers who arrived by car. Streets were widened and resurfaced in brick or concrete pavement. They were provided with traffic signs and with painted stripes along the curbs marking spaces where cars could be parked in orderly fashion. Improved street lighting enhanced the look of business areas at night by better illuminating streets, sidewalks, and adjacent building facades, including their signs.

Storefront improvement was another way of increasing Main Street's commercial effectiveness. There were basically three ways to proceed. The simplest strategy was to adopt new signage, the signs themselves treated merely as appliqué. Traditionally, signs were attached to building facades very much after the fact of building design and construction. They were, in other words, rarely considered part of a building's overall design, a reflection of the fact that most main street storefronts were built on speculation to be rented out over time to a sequence of merchants, all of whom would bring their own signage. A second strategy was to enlarge or otherwise improve display windows, coordinating their improvement, in turn, with new signage. A third strategy was to totally redo a storefront, fully integrating new signs into the design as part of an architectural motif.

REPLACEMENT SIGNS

Most Main Streets were built to a standard template. Typically, buildings were erected abutting one another on deep lots of some standard dimension, large buildings occupying two or more lots. Only facades were finished with ornament, quality materials and fine workmanship being fully evident there. In their alignment along a street, buildings stood not so much as three-dimensional structures, but as variously decorated wall planes, buildings on corner lots excepted. Main Street buildings displayed remarkably little variation, there being relatively few building types nationwide.[7] The least impressive structures were single-story, although frequently a "false front" (that totally obscured the roof behind) gave the impression of greater height. More impressive were two-, three-, and four-story buildings, their facades usually divided into two distinct horizontal zones: a one-story lower section, clearly public in orientation, and an upper zone, more clearly private. Taller buildings, on the other hand, tended to a vertical three-part organization. On a base, articulated as retail space, an upper array of five, six, seven, or more floors rose to be capped by a clearly visible roof or exaggerated eave or cornice. Such buildings, rather than appearing fully integrated into "building blocks," tended, for their size and styling, to stand apart in visually anchoring a street as in Casper, Wyoming (fig. 9).[8]

Most commercial signs logically emphasized lower-level storefronts. There was an amazingly diverse array of signs to be variously displayed on commercial buildings. Witness Schlanser's Delicatessen (fig. 10). Fascia signs include a lintel or overhead sign (at top), column signs (left, center, and right), and a sill sign (lower right). Also pictured is gold window lettering, and, inside the display window, a shelf sign. A placard on the door gives the store's opening and closing hours. The exterior signs are all of varnished galvanized metal and are attached to the building with underlying wooden frames.[9] Lacquered fascia signs dominated Main Street storefronts until enameled-porcelain and acrylic resin plastics were popularized in the 1930s.[10]

As in big city downtowns, changing sign fads and fashions prompted both frequent sign replacement and the constant complementing of old signs with new. Certainly, new signs were required when new businesses were started. But important also was the Main Street merchant's increasingly closer ties with national corporations and their trademarked or branded products and services. With increased intensity through the 1920s, large retail corporations opened branch or franchised stores in small town business districts. In the grocery trade, the Great Atlantic and Pacific Tea Company led the way with some sixteen thousand branches operating nationwide by 1927, with regional giants Kroger and Piggly Wiggly operating some four thousand and twenty-six hundred stores, respectively.[11] Chain stores could offer lower prices through economies of scale inherent in mass purchasing and through streamlined distribution. Also, A&P, and many of the other food store chains, integrated back into manufacturing, bringing to their stores, without intervention of wholesale jobbers, their own branded merchandise. To compete, many independent grocers affiliated in voluntary chains and thus developed their own branded products. Organized in 1926, for example, was the Independent Grocers Alliance, participating stores all displaying the IGA logo.

FIGURE 9

Center Street in Casper, Wyoming, c. 1935. Larger buildings, with six, seven, or eight office floors stacked above ground-floor retailing, anchor the street visually.

FIGURE 10
Schlanser's Delicatessen with a diversity of signs. Source: "Bids on Signs for Entire Store Front from Maine to California," Signs of the Times *50 (July 1925), p. 82.*

At work at chain store locations was a form of place-product-packaging: a marketing device that involved creating look-alike sales outlets all programmed more or less the same. In gasoline retailing, each gas station in a chain was similarly structured and decorated, both inside and out, signage being fully standardized. The same goods and services, all priced the same, were made available at every location, customers knowing exactly what to expect.[12] At storefront grocery stores, however, "sameness" was primarily a matter of signage, the building itself often being a rental space, a generic storefront on Main Street.

Few corporations, whatever the industry, sought to own outright the retail establishments that carried their products. It was simply too costly to do so. Even the automobile makers resisted the temptation to own stores, developing, instead, networks of licensed dealers to handle car sales within protected trade territories. Local merchants, in turn, found that by associating themselves through licensing or franchising, or even through mere display of a well-known corporate logo, local sales could be enhanced. This was especially true once national advertising campaigns, focused especially in newspapers and magazines, brought brand-consciousness fully to the fore.[13] Local merchants shared in the fruits of brand loyalty while, at the same time, retaining degrees of independence. Also, branded merchandise greatly facilitated the rise of self-service which, in turn, cut labor costs.

Of all the corporate logos to influence the American scene early in the twentieth century, none became so widespread so quickly as that of Coca-Cola. The logo first appeared on large painted wall signs and then on signs hung over store entrances. Porcelain-enameled signs with white letters spelling out "Drink Coca-Cola" on a red background (with "Fountain Service" in yellow lettering on green across the top) were introduced in the early 1930s and became the single most important cue to soda fountain location in the country (fig. 11). *Signs of the Times* editorialized: "The repetition of these emblems at the point of purchase keeps buyers constantly reminded of convenient sources of supply and increases the value of all advertising used to develop buyer consciousness of the product, its usefulness, and the desire of its possession."[14]

With the end of Prohibition, breweries across the nation similarly emphasized point-of-purchase signage. Tavern owners, in agreeing to feature a company's beer, could obtain a wide array of advertising displays. Outside, an electric sign might be hung over a tavern's door, a fascia sign placed across its facade, or signs stenciled or painted on its window glass. Inside, smaller electric signs might be hung on walls, along with such accoutrements as clocks, light fixtures, and calendars, all carrying a company's logo. In 1934 the Joseph Schlitz Brewing Company ordered 2,150 double-faced, luminous-tubing, porcelain-enameled outside signs; 1,200 box-type porcelain-enameled window signs; 100 double-faced, luminous-tubing dealer-name signs; and 200 "skeleton" signs, all intended to reinforce dealer locations. Manufactured by the Flashtric Sign Works of Chicago, the order involved some 258,500 feet (50 miles) of neon tubing, 38,850 feet (7 miles) of cable, 41,000 electrodes and bushings, 91,000 square feet of sheet steel, and 5,800 transformers.[15]

As corporate logos proliferated on Main Street signs, store owners enjoyed connection with recognized products, products that carried for customers implications of quality mass-produced and highly standardized. Corporate logos made a local retailer appear up-to-date and even modern, tying a store into a world of consumption no longer merely local or parochial, but urbane if not cosmopolitan.[16] Increasingly apparent on

FIGURE 11
Coca-Cola sign in Sebree, Kentucky, 1990. Notice the competing Pepsi-Cola machine (with its point-of-purchase signs) being patronized on the sidewalk below.

America's Main Streets were the signs of the new automobile age, the automobile dealership having become a ubiquitous presence. General Motors introduced neon "identification signs" for its Buick dealers in the 1930s. These offered a symbol, commented one journalist, "that the Buick owners and users will come to look for when they require service for their cars, as naturally as they look for the identifying signs when they want to purchase their favorite gasoline."[17]

Retailers, however, were usually required to either buy or lease such signs. Sponsoring corporations did share costs, in a sense, in that they subsidized sign design and, in placing large sign orders, reduced a merchant's purchase outlay through economies of scale. Initially, public utility companies operated departments or owned subsidiaries that solicited storefront sign business. Power companies typically encouraged store owners to pay for their signs on a monthly pro-rated basis in power bills. In the mid-1930s, dealer identification signs typically sold for about $1,000, or some 60 percent of a merchant's rental costs over a 36-month period.[18] In 1935 alone, and despite the economic doldrums of the lingering Depression, some thirty-four hundred sign manufacturers sold an estimated 1,100,000 signs worth an estimated $205,000,000. The electric sign industry employed some fifty-seven thousand people in shop, office, and sales work.[19]

DISPLAY WINDOWS

As in big city downtowns, display windows were important in bringing sign art to America's Main Streets. As storefront signs attracted attention, so also did window displays. But, more importantly, display windows could also inform and persuade as they allowed shoppers to actually see examples of what a store had to offer. Shopping implies personal gratification directed toward self-empowerment. It is a testing of self that pits "just looking" against actually having (fig. 12). "To shop," observed social theorist Anne Friedberg, "is to muse in the contemplative mode, an activity that combines diversion, self-gratification, expertise, and physical activity."[20] With the consumer revolution came shopping as a kind of leisure-time pursuit.

Display window advertising always involved signs as well as things displayed. Signs in display windows, just as with storefront identification signs, increasingly coordinated with advertising in newspapers, magazines, and other media. Display windows needed to attract attention, but they also needed to arouse interest, win confidence, and create or reinforce the desire and disposition to buy then and there. Through World War I, most small town merchants, as with their counterparts in

FIGURE 12

An ink-blotter advertisement for Everett, Aughenbaugh and Co., 1933. Shopping behavior was learned early in America, where people were daily encouraged to want what they didn't yet have, as through the temptations of a bakery display window.

the nation's big cities, crowded their display windows with merchandise. Displays stood as comprehensive summaries of what stores contained within, while signs on and in windows tended to emphasize primarily quality and price (fig. 13).

After World War I, highly simplified window displays were widely popularized, organized around the showing of only a few items of merchandise. Often, windows did not contain merchandise at all, but merely something attention-getting around which sales might be pitched. Mainly, however, the idea was to place examples of a store's offering in some stylish or useful context. One did this by manipulating background such that important social meaning (especially implications of class and status) might attach to the items displayed. Displays were also intended to make the store itself appear desirable as a fashionable place to patronize. It was not just what one bought, but where one bought it. Association with place of purchase, and not just the things bought, loomed increasingly important in defining life style. Less cluttered, the display window's three-dimensional potential for the dramatic quickly became apparent, especially when the use of mannequins for displaying clothing came to the fore in the 1920s.[21] Increasingly important were such visual abstractions as color harmony, compositional equilibrium, and contrasting textures.[22] Window trim-

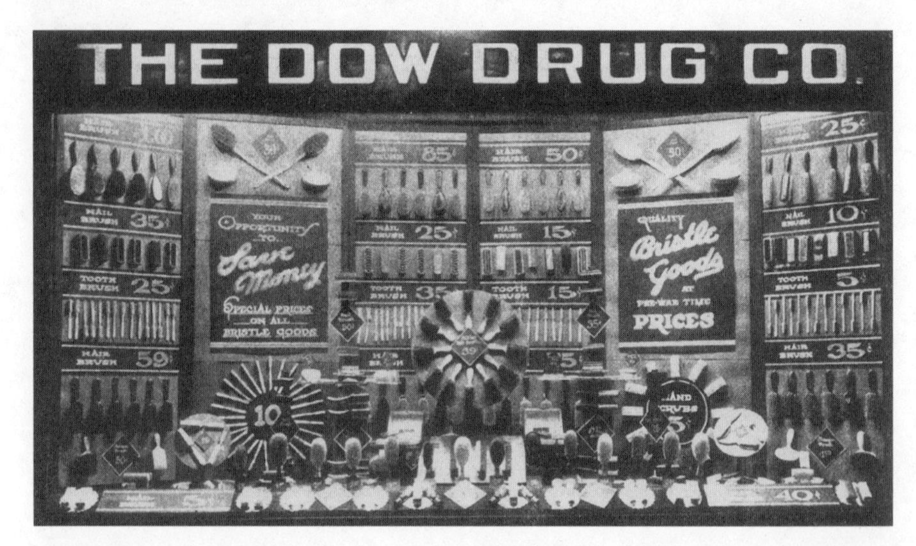

FIGURE 13

Drug store window typical of the immediate pre–World War I era. Source: Edward Hugill, "The Missing Link in Successful Merchandising: Making Window Display the Golden Link," Signs of the Times *31 (Feb. 1919), p. 10.*

ming became a profession, aided by such organizations as the National Association of Window Trimmers of America, the National Association of Display Industries, and the National Association of Specialty Advertising Salesmen.

Display window signs were of two basic types: those embossed or otherwise attached to display window glass and those simply displayed behind the glass, prime among these being so-called showcards. Of the former, painted lettering on window glass and transparent sign appliqués remained popular through the 1940s. Of the latter, one-of-a-kind lettered placards were originally placed inside display windows to complement displayed items. Most were hand painted, the air-brush greatly facilitating their production in local sign studios. But lettering machines, employed mainly by department stores, enabled quick production of showcards in at least small batches. When large companies distributed sizable numbers to retailers across a region or the nation, then placards were invariably mass produced using photomechanical and other printing techniques (fig. 14).

STOREFRONT REMODELING

Beginning in the years before World War I, prism glass was set in window casements positioned across building facades, usually just above display windows (fig. 15). Their purpose was to allow more natural light to flood inside. At the same time, metal window frames were much less bulky than wooden frames, making display windows appear larger.

In the 1930s, however, even more dramatic change came to America's Main Streets. An amendment to the National Housing Act of 1935 encouraged storefront remodeling as a means of pump-priming the still depressed economy. Remodeling loans up to $50,000 were partially insured by the federal government. "Modernism" quickly became the order of the day, a design imperative that emphasized smooth, clean, functional surfaces.[22] Streamlined forms with sweeping, curvilinear lines became very much the fashion. Fascia signs, rather than being applied to facades, were made integral to them. Indeed, storefronts became giant signs in their articulation. However, renovation was largely reserved to ground floors in multistory buildings. These new signlike fronts communicated less with pedestrians than with motorists. Remodeled storefronts usually stood in stark contrast with remaining traditional facades. Whereas previously most Main Street facades appeared to be more or less integrated visually — sharing as

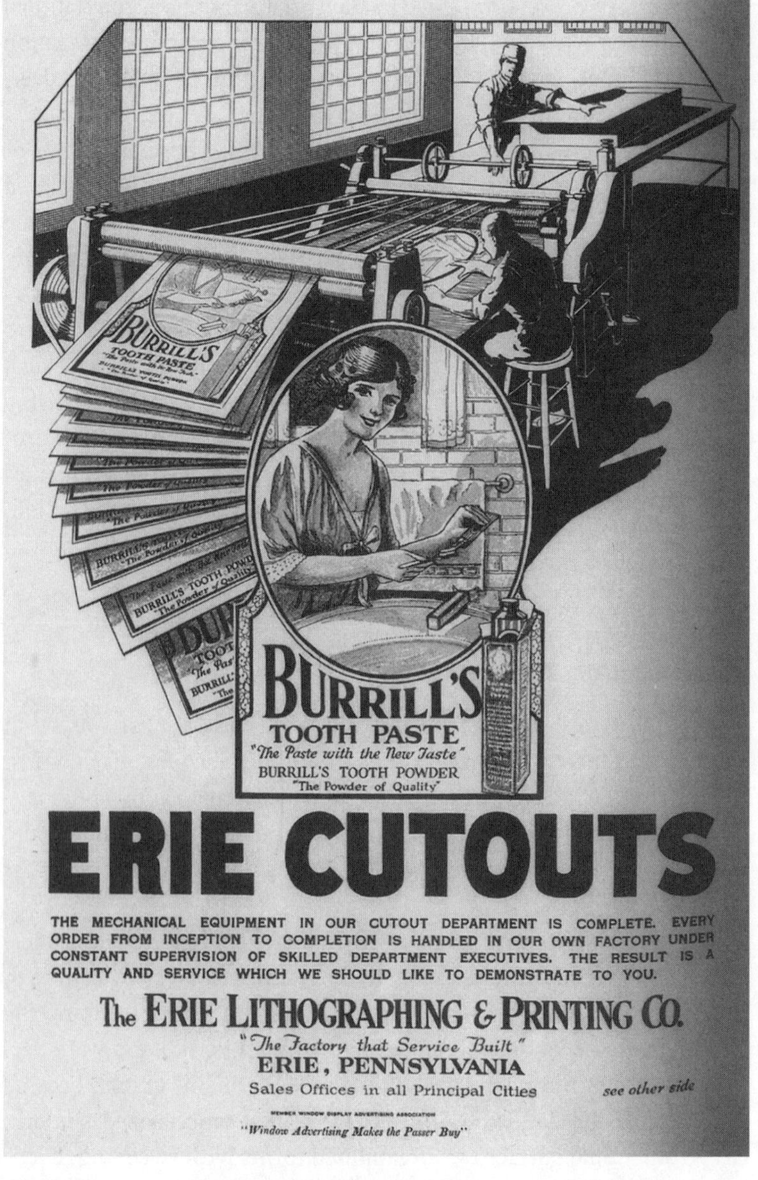

FIGURE 14
Advertisement for the Erie Lithographing and Printing Company. Source: Poster *16 (Oct. 1925), p. 34.*

they did traditional architectural styling, traditional building materials, and traditional signs — remodeled storefronts now deliberately stood out, tradition largely discarded. With the use of opaque structural glass, porcelain enamel, glass block, glazed brick, stainless steel, and alu-

FIGURE 15
Prototype display window manufactured by the Pittsburgh Reflector and Illuminating Company. Source: Signs of the Times *23 (Aug. 15, 1913), p. 24.*

minum screening, among other materials, Main Street stores were made shiny and light-reflective, especially at night when floodlit or enhanced through neon signage.[23] Increasingly dominant after World War II were signs fully sheathed in translucent acrylic resin plastic and backlit with fluorescent tubes (fig. 16).

Along with storefront renovation, Main Street merchants also undertook to modernize store interiors, rearranging store layouts to more fully expose merchandise to view, thus encouraging self-service. Many merchants even began to eliminate their display windows, a trend that accelerated quickly after World War II. Instead of viewing window displays, pedestrians and, more importantly, motorists, could glimpse a store's full interior. They could clearly see its self-service implications, if not the merchandise itself, displayed on counters. With the new fluorescent light fixtures, store interiors, especially at night, shone brightly, very much like large signs or, better still, given the tenor of the times, like giant television screens. So-called visual or open fronts came to figure most prominently in the new shopping centers that anchored peripheral commercial strips in the larger towns. Indeed, the open storefront was one means by which traditional Main Street merchants

FIGURE 16
Design for a signlike storefront backlit at night. After dark, translucent materials could be lit from behind to make a building facade literally shine like a sign. Source: Signs of the Times *72 (Sept. 1932), p. 39.*

sought to emulate, and thus compete with, new automobile-convenient stores out along small town America's new commercial strips.

Main Street storefront remodeling continued through the late twentieth century, but, given the lack of vigorous economic growth in most small towns, renovation tended more to the ad hoc and the superficial, much of it merely cosmetic. Embraced were inexpensive facade claddings, such as aluminum, acrylic plastic, or even wood board. Sign lettering tended to be plastic and generic. Entire building facades were sometimes sheathed in cheap "cover-ups" to hide vintage ornamentation or dated building materials.[24] Thus, a makeshift look was brought to America's Main Streets.

Beginning in the 1980s, lettered canopies or awnings, brightly backlit by fluorescent tubes, rapidly came into popularity, another means of inexpensively updating traditional storefronts.[25] Backlit canopies were especially popular in localities with strict sign ordinances, regulations placed on regular signs but not applicable to canopies or awnings.[26] With new canvas fabrics developed, "flexible sign faces" came to be widely used in cheap replacements for traditional storefront signs. Very important to the use of flexible sign material were digital-imaging and computerized screen printing.[27]

■ Essentially a pedestrian place to begin with, the quintessential small town Main Street became increasingly automobile-oriented, a change clearly reflected in storefront signs. From identification signs projected out over sidewalks to the backlit awnings or canopies, visibility from passing cars evolved as the dominant consideration. Storefront remodeling first anticipated and then emulated architectural ideas fully developed along the nation's roadsides. On Main Street, open storefronts did away with display windows to allow motorists, and not just pedestrians, to look inside stores. At night, open storefronts were lit like giant signs.

Signs denoted. They informed, outlining for potential customers what Main Street businesses were all about. Communicating a store's name, or the name of its proprietor, and the goods and services that a store provided was a sign's main purpose. But signs were also connotative. They symbolized, adding meaning through implications of size, design, brightness, or other characteristics. Signs were there to be read, but they enabled even wider reading of landscape as they visually and functionally played off other elements of the built environment. Perhaps most telling was the merger of sign and building facade following the dictates of modernism. Symbolized was modernity, but so also, in the context of the times, faith in the future, especially a future linked to automobility.

Small town Main Streets, like big city downtowns, were more than just business concentrations. It was there that the emerging consumerist society was first focused. Americans, small-town people included, increasingly came to define personal and group identity in terms of the goods and services they consumed. Community in America took on decidedly materialistic implications. Point-of-purchase advertising encouraged Americans to consummate their life-style choices by acting on their shopping impulses. In the competition for customer attention, storefront signs also offered alternatives. By doing so, they seemed to empower the individual.

3 : Roadside Signs

Symbolically and physically, the automotive highway with its travelers' largely commercial services beside it emerged as the newest and most influential landscape in America as the twentieth century unfolded. Highways and their car, truck, bus, and motorcycle drivers and passengers founded a new means of expressing the American zeal for movement, whether unflinchingly for work or blithely for recreation. It took more than the availability of highways for Americans to achieve the comparative novelty of free passage over so large and varied an expanse as a virtually continent-wide nation. The social and economic conditions enabling America's car culture are not exceptional, but they are remarkable. What stands unprecedented is the depth to which most Americans generally committed unquestioningly to car culture, including buying and maintaining their vehicles and paying taxes for the travel infrastructure — the combination of vehicles and roads giving birth to a car-convenient landscape. The whole of this novel American landscape comprises not only the vehicles coursing its pavements (made especially for them), those pavements themselves, and the dedicated right-of-way services, but also the signs directing traffic and the signs advertising goods and services. The history of the sign industry in the twentieth century is virtually synonymous with the growth of sales that signs arouse in automobile-borne travelers. The highway has been explained as its own habitat, with people taking special roles and a distinctively fashioned material culture, but signs have been the least systematically treated contributors to this highway habitat.[1]

THE DAWN OF ROADSIDE COMMERCIAL SIGNS

Evidence points to wide human dependence on roadside signs, and not only those persons wealthy enough to travel by car. Hobos in the Depression drew ephemeral symbols in coal or gypsum on utility poles to tell their "brothers of the road" of pleasing or troublesome places ahead.[2] Advertising companies leave the most lasting and noticeable signs on the landscape. We next look at how outdoor advertising leaders restructured their industry with the coming of the automobile.

Selling its services: that primary objective compelled the outdoor advertising industry from its beginning. Indeed, it has remained so.[3] In fact, the industry has experienced periodic spans of increased and decreased demand, although to the general observer outdoor advertising has seemed a given in contemporary life. That status of being taken for granted attests to the entrepreneurs' resilient and creative persistence but no less to a broadly based cultural dependence on advertising for what and how Americans consume. Historian Gary Cross, surveying economies globally at the end of the twentieth century, demurred about the widespread belief that capitalist-inspired consumerism had won and was the only future economy. He also answered more persuasively than any other scholar why consumerism had prevailed to this time. It meshed with America's hegemonic values of liberty and democracy, giving the chance to find one's individual identity without forfeiting cherished family, friends, and culture.[4] Consumerism also eased class tensions, although it often exacerbated gender and race divisions for immediate financial benefit. And advertising signs were a mainspring of that consumer culture. They would probably be in place without frenetic sales campaigns for them but likely not in the same numbers, of quite the same content, or in the same locations without the very self-conscious outdoor advertising industry.

During the Depression of the 1930s, when outdoor advertising entrepreneurs initially confronted decreasing demands for their services for the first time in several decades, a sense of urgency swept their ranks. To lure back advertisers, the industry's leaders set out to prove the effectiveness of their media with verifiable data. Circulation — in common parlance, the number of viewers, whether in cars, on foot, or on mass transit, passing a sign within a given time — became critical to the industry's strategies in the 1930s. Circulation had been considered earlier, but assessments frequently had been informal and taken only for big signs. The Barney Link Fund, named for one of the industry's most active entrepreneurs in the early twentieth century, had supported circulation research at the University of Wisconsin since 1924. In 1931 the Association of National Advertisers' outdoor committee, chaired by an executive of a big advertiser, Coca-Cola, pushed for reliable studies. In that same year, the Outdoor Advertising Association of America voted funds to conduct more precise research on circulation. Miller McClintock, renowned head of the Bureau of Street Traffic Research at Harvard University and an active consultant in the new field of traffic engineering, supervised the project. The attraction to McClintock's authority on research into traffic congestion underscores the outdoor advertisers' reliance on customers in automobiles. The Traffic Audit Bureau, which the industry established in 1934, produced data permitting advertisers to know the actual number of a sign's viewers. It still operates. Measuring the efficacy of signs became a behavioral science. Kerwin H. Fulton, chief executive first in the Poster Advertising Company in New York City and a primary force in the creation of the General Outdoor Advertising Company in 1925, founded Outdoor Advertising Incorporated in 1931 to spur the concept of the outdoor medium as advertising. The tobacco industry, long the outdoor medium's staple, had shifted accounts to the novel radio and put the outdoor medium's efficacy into question. The various remedies effected improvements throughout the Depression, profits increasing from $20,000,000 in 1932 to $36,370,000 in 1939, and advertisers increasing from 352 to 520 in the same years. Many merchants finally abandoned their lingering doubt that signs embodied true salesmanship. Evaluating signs' costs relative to circulation, they came to appreciate signs in terms of their efficiency.[5] The course of forming a public and selling advertising to direct that public matured the industry.

What led to this juncture? Before the right-of-way beside the automobile highway became the outdoor advertising industry's most influential and lucrative seedbed, the industry had taken root and instinctively developed its material culture along city streets. Most of the advertising signs bore little resemblance to those fashioned expressly for the roadside, and their location differed markedly from those placed for viewing from automobiles. Once signs were designed and posted, retailers were satisfied. Neither the signmen nor the merchants paid much attention to the signs' location.[6] Often advertisers of different products clustered those signs on a single pole or fence, producing a visual hodgepodge. Visual confusion hobbled such advertisements (fig. 17). Competition for attention more often resulted than any single sight line focusing vision. In style, city signs antedating the automobile were often comparatively small, for example, three by five feet, and busy with words characterizing the advertised commodity (fig. 18).

Other sign types, both large and small, were made available to advertisers. Point-of-sale signs included sidewalk stands, signs hanging from poles or display standards, and hanging signs. Each customarily displayed shop names or brand logos. Wall signs defined the larger sign sizes, averaging four hundred to fifteen hundred square feet, and often were off the premises of the businesses where the advertised commodities were sold. Automobile accessory dealers, garages, and petroleum dealers, all dependent on automobilists, innovated continually to attract customers. A new Standard Oil station built in 1926 at

Willoughby, Ohio, was far less impressive than its electric spectacular, perhaps the largest for a gas station until then, built by the Halter and Ragg Sign Company of Cleveland. "Red Crown" in letters thirty-six inches high and "Gasoline" in letters thirty inches high stood out on a sign twenty-five feet wide and thirty-two feet high.[7] Eventually the proliferation of gas stations alone, not to mention other advertisers, turned city streets into sign corridors.

"Sniping," the practice of posting small signs on fences or trees and often without consent of the property owner, characterized much of the advertising by open countryside roads as late as the 1890s, on the eve of the automobile's appearance. Indeed, sniping was so common that people distinguished between "daubers," painted snipe signs, and "tackers," printed snipe signs. In 1925, when the Outdoor Advertising Association of America adopted standards against any member daubing or tacking, the practices, already in decline because of local ordinances against them, substantially disappeared.[8] In those comparatively rare instances where they cropped up in later times, including those painted

FIGURE 17

Louis Schwartz's Highway Displays, Inc., in Poughkeepsie, New York, 1907, typifies some of the unpretentious bill posting for early travelers. A seventeen-year-old entrepreneur at this time, Schwartz built his company until, forty years later, it served highways in Connecticut, New Jersey, and New York. Source: Signs of the Times *150 (Jan. 1947), p. 32.*

FIGURE 18

Roadside advertising still calls forth simple salesmanship in many places, as this 1999 photo of a roadside farm stand near Porter, Indiana, demonstrates.

on locations with the owner's permission, their reduced numbers often made them pleasing novelties. Mail Pouch tobacco and "See Rock City" were notable examples (fig. 19).

As the outdoor advertising industry began to target sign styles and locations for automobilists, this further spurred the trends in signs already under way: emphasizing images over words, psychologically analyzing consumers, and posting signs less frequently but more thoughtfully for each advertising campaign. Impetus for change came from within the outdoor advertising industry. Consider automobile advertising signs themselves. Studebaker led the auto industry in advertising in 1915 but relied on local poster plant owners to persuade local Studebaker dealers to buy outdoor advertising space.[9] *Signs of the Times* editors stated in 1913 that "outdoor advertising men should begin at once to know and educate the local [automobile] dealers to know and appreciate the power of outdoor advertising in boosting his particular business." Outdoor advertisers rapidly appreciated the young automobile industry's use of market segmentation along social lines — a particular marquee for a particular group of customers — and favorably contrasted outdoor advertising with the printed media, noting that signs that targeted a neighborhood provide more effective circulation than ads in a newspaper, which often has a wide but varied readership. Claiming for outdoor advertising the advantage over newspapers of

FIGURE 19
Real estate developer Garnet Carter and his wife, Frieda, parlayed Rock City near Chattanooga, Tennessee, into one of America's best publicized roadside attractions with barns bearing hand-painted signs. This example stood near Palmyra, Indiana, in 1978.

sign size and color, as well as strategic location, *Signs of the Times*'s editors summed up that outdoor advertising had "punch."[10]

From the wide variety of sign types born in the city, the billboard became the one upon which outdoor advertisers concentrated. Billboards especially lent themselves to big, bold, and less easily avoided appeals than the print media in a culture of increasing automobility. Distinguishing outdoor advertising from newspapers, Kerwin H. Fulton, writing from long experience in advertising, explained the difference between the cognitive versus the affective essence of the two media: "Newspaper advertising gives the logic and the argument ('reason why,' if you like), and poster advertising contributes the powerful suggestive value of color — a picture — and three or four words pregnant with meaning."[11] Poster designers instilled those qualities in common posters before the automobile, to be sure, but billboard designers for roadside advertising worked to refine those commercial virtues within artistic sensibilities.[12] They wanted to attract, not stun, the viewer.[13]

Billboard structure changed when advertisers determined that advertising could play a permanent role in commerce. Fred Ruby of the Ruby Manufacturing Company in Jackson, Michigan, perfected a construction technique that solved two of the billposter's persistent problems: painting outdoors, which made painting posters dependent on clement weather, and billboard construction, where what was needed were sturdy billboards easily assembled.[14] The industry also sought — and found — ways to mass produce signs at low cost and rapidly relocate them in response to shifting transportation patterns. Elements of the contemporary roadside billboard came to the fore quickly.

Railroad corridors had been a harbinger of how advertising aimed at automobile occupants would transform the use and look of automobile rights-of-way. The automobile's forerunners in mass transportation, trolleys and railroads, accustomed intracity riders, commuters, and long-distance travelers alike to posted advertisements. Electric streetcars fostered windowside viewing beginning in the late nineteenth century, when public transportation ended the pedestrian-bound city with its general radius of two to three miles around the city center. Developers speculated that real estate parcels along streetcar lines bought for greater income in the future could earn income immediately through shops selling essential commodities to surrounding neighbors, who thereby avoided the inconvenience of travel downtown to the old urban centers. "Taxpayer" strips, so called because they paid their owners' present taxes, developed across the nation. Empty lots afforded outdoor advertising the chance for wall signs and billboards (fig. 20). Speculating beyond existing cities, real estate developers built suburban settlements and streetcar lines to them. The routes between the old centers, where people yet traveled to work, and the new suburbs, where they lived, multiplied opportunities for advertising displays. The visual consequences outstripped the capacity to understand what signs in the landscape meant by looking at them in localized displays at or near the point of sale. Advertisers imagined far-reaching empire (fig. 21).[15]

Steam railroads stretched the advertising corridors even further. Before the number of railroad passengers peaked in 1920, large billboards (ten feet high by fifty feet long and twenty feet high by one hundred and fifty feet long) arranged parallel to the tracks formed advertising corridors similar to those automobilists began encountering in the 1920s. The most extensive railroad display by the early 1920s linked boards one thousand two hundred feet in length in New Jersey for the Wrigley chewing gum company. In contrast, Wrigley arranged two hundred-foot long boards along automobile highways.[16] Emphasis soon shifted in favor of the automobile travelers.

Automobile highway extent, a key permissive factor in roadside sign development, leaped in miles beginning in the 1920s.[17] From its

FIGURE 20

Walker and Company of Detroit, one of the nation's large early billposting services, here demonstrates how billboards appeared in the neighborhoods along radiating streetcar lines. Source: Wilmot Lippincott, Outdoor Advertising, *p. 205.*

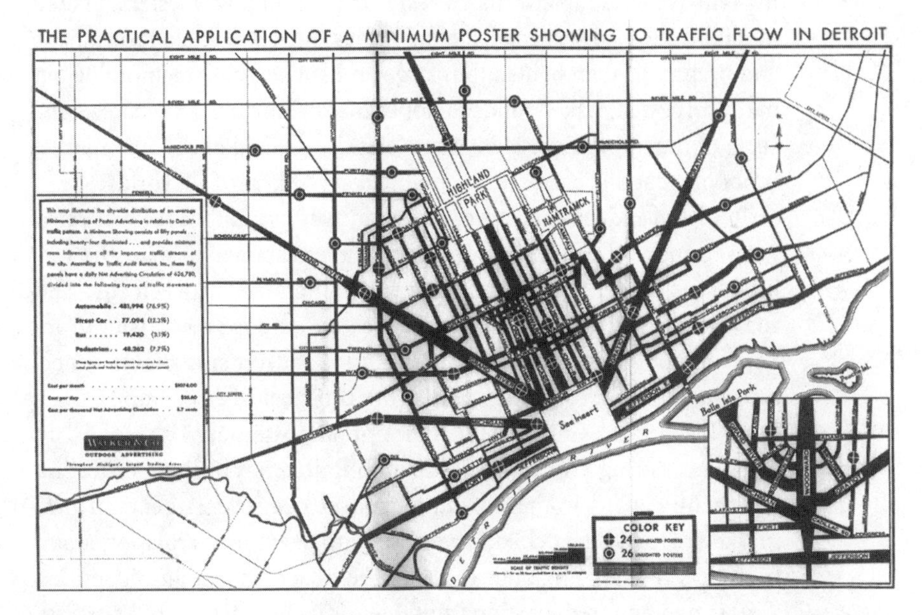

FIGURE 21

Walker and Company (one of whose characteristic billboards is shown in fig. 20) mapped its displays in this 1930s example to convince would-be advertisers how a daily circulation of 626,780 was achieved. Source: Hugh E. Agnew, Outdoor Advertising, *facing p. 116.*

vantage point in San Francisco, Foster and Kleiser succinctly declared in 1923:

> There is a broad white band of concrete roadway, extending over 1300 miles from the Canadian border to Mexico, traversing the States of California, Oregon, and Washington. This is the "Main Street" of the Pacific Coast, along which have been and are being built prosperous, active cities and towns.

"Posting 'Main Street'": Foster and Kleiser's slogan, said what other outdoor advertising companies were doing, as advertising extended from downtown into the countryside.[18]

Thus, the 1920s became a watershed in roadside sign development. Previously unimagined highway expansion opened ever more and longer roadsides for billposting. Outdoor advertising entrepreneurs thought of winning larger contracts from advertisers and diminishing public disapproval through self-restraint, all the while substituting strategic sign location for the traditional saturation. Aggressive billposters instinctively correlated business potential with enhanced public infrastructure. No published or spoken clarion calls prefigured the rising ranks of outdoor advertisers beside the new highway system. Like so much of the road and roadside industries, innumerable opportunistic and pragmatic merchants, both large and small, some memorable but many forgotten, risked funds in hopes of "making it big," whatever the scale of their ambition. Here was a classic seedbed of the free-market economy that those people championed as the best of all worlds.

Regional and national advertising campaigns induced sophisticated ways to more effectively advertise and take responsibility for the way billboards looked and their sites were maintained. Between 1917 and 1927 advertisers increased expenditures on outdoor advertising from $2,000,000 to $85,000,000. Earlier outdoor advertising plants generated business for local merchants, the prospect of circulation in far-flung locations seeming visionary if not preposterous. The Hood Tire Company during World War I was among the first businesses testing and trying ways to profit from outdoor advertising over larger than local markets. Hood dispatched survey crews to examine circulation along highways in New England, mapped the observations, and identified almost two hundred locations suitable for its prospective bill posting. The scheme aimed at calculating the number sufficient to carry the message effectively and position the signs to maximum advantage. Some of these sites were calculated to promote driver safety by drawing attention to dangerous curves and other hazards.[19] Burma Shave demonstrated how small signs, too, could be turned to admirable effect. Begun in 1925 in Minnesota, when the construction of the trunk highway system had just gotten under way, the humorous doggerel of the Burma Shave signs (which were placed in a series spaced at eighteen-second intervals for automobilists traveling at 35 miles per hour) became a fixture of Americana before they were taken out of use in 1963. Only four states prohibited Burma Shave signs, and the signs were discontinued only because speed limits on the newer highways made them unreadable. Travelers had gleefully sought relief from wearisome trips by reading the latest Burma Shave jingles.[20]

Companies budgeting big advertising campaigns, principally national or regional showings, went far toward redefining outdoor advertising as a medium of the automotive roadside. By the late 1920s, the outdoor advertising industry's argot distinguished between signs according to location, namely, "suburban bulletins" and "highway bulletins." The former were posted, according to the Outdoor Advertising Association of America's official lexicon, "within the populated area of a city, along or at the intersection of the important automobile boulevards leading to important urban centers and at out-of-town points." The latter were posted on highways between population centers and, before highway intersections spawned various combinations of tourist courts, gas stations, and restaurants, gave mileage and directions to nearby traveler services.[21]

Entrepreneurs pushed the material culture of the billboards to exceptional excess before settling on simplified forms. Foster and Kleiser hoped that framing billboards with columns topped by fully sculptured female heads and busts would bring respectability to outdoor advertising via allusion to classical art. These billboards, named "pilasters" — the trade preferred the name "lizzies" — experienced popularity only on the West Coast and only in the 1920s because they were too difficult to maintain. Formal gardens planted at the base of those billboards added a garden aspect to clusters of billboards called "poster parks" and accentuated the sense of a place contrary to sales hustle (fig. 22). Suburban pastoralism, long a popular trope in American culture, suffused the dark-green–painted lattice aprons that became common throughout the nation in company with the twenty-four–sheet poster panels that first appeared beside streets and highways in 1925 and 1926. Thereafter, steel frames and lattices replaced their wooden predecessors, and the common billboard's components through the 1940s were set.[22]

Innovations usually incorporated designs and technologies calculated for the greatest desired effect at the least cost. Most commonly,

FIGURE 22
"Lizzies" behind a garden, characteristic of this billboard type.
Source: Signs of the Times, *93 (May 1956), p. 125.*

highway signs were synchronized with point-of-purchase signs, making the new highways out of towns into extensions of the cities. Older tactics, such as direct mail to prospective customers, proved less effective in selling than did the combination of country highway and city street point-of-purchase signs. Salesmen deemed advertising urgent during the Depression to boost consumer demand but seldom could afford more than modest changes. Small signs were pressed into use. Champion Spark Plugs, for example, erected twenty-five hundred galvanized steel signs with faces on each side for the first national campaign using light-reflecting buttons, 370 on each sign. Reflector buttons, a cheap but effective technology given the increased amount of night driving, had been introduced a few years before but gained wide use two years after the Champion signs in 1933. Architectural modernism's faith in machinelike modeling encouraged some designers to craft buildings imbedding corporate identities and project brands as images.[23]

Roadside advertising became the venue of choice among innovative advertisers with larger budgets by midcentury. Small businesses employed low-cost and traditional techniques. For example, "exploitation films," which occupied the niche between visionary art films and

Hollywood fantasies, reached the height of their popularity in the 1930s and 1940s with the promotional work of "roadshowmen." The exploitation genre's gritty portrayal of stories about such topics as drug abuse and prostitution filled the few surviving independent movie houses after roadshowmen traveled in advance of a showing to arrange terms with the theater manager, including display of colorful posters and, for shop merchants, window cards.[24]

THE ROADSIDE SIGN INDUSTRY IN THE LAST HALF-CENTURY

A spate of little signs and billboards on highway shoulders throughout the open countryside represented the most extensive appearance of new signs nationwide through the 1930s. The mutation of roadside buildings into signs beginning in the 1920s and 1930s (fig. 23), however, generated a new type of landscape — the strip.

Commercialized transportation corridors manifested several immature forms. In large urban centers, wall signs were displayed high on buildings for windowside viewing from elevated trains. Later, in cities such as Detroit in the 1940s, when streetcars gave way to automobiles, buildings of ebbing economic vitality retained their use almost exclusively for billboards mounted atop them. Here the sign's scaffolding reduced the aesthetic appeal of the host building. By the end of the twentieth century, on interstates threaded tightly through closely packed old street grids, such as in St. Louis, pole signs helped create a tiered landscape. Rising high above street level, pole signs were mounted just above the elevated interstates at various points to be viewed by those affluent enough to travel by car, while, on the streets below, marginal shops were left to their once strategic street corner locations and small on-premises signs to snare attention. In those cases, signs surpass buildings altogether for communication. Speed has rendered symbolism expedient.

Critics in the 1920s first verbalized the roadside transformation. Multiplying roadside mileage engendered a growing display of billboards and buildings. Critics perceived billboards as too many and indiscriminately placed. Maine, for example, contradicted its program to attract tourists by letting billboards encroach on the scenic coastline between Portland and the state line and was better called "Billboardia," quipped one disapproving journalist.[25] Billboards intermingled with unusual architecture set off a backlash. Some of the new roadside buildings seemed freakish, contrary to elite tastes, and their grounds were often deemed ill-kempt according to standards of housekeeping

FIGURE 23
On one site, Windy Acres Motel and Café at Cannon Falls, Minnesota, typifies the transformation of signs from free-standing units (left) and key identifiers of the services on buildings to a group signaling an oasis for automobile travelers. Units like this strung side by side grew into a strip.

promulgated since the Progressive era. Some of the freakish architecture was mimetic — hot dog stands shaped like hot dogs, auto courts with cabins shaped like tepees, for example. Other was oversize — Paul Bunyan and his blue ox, Babe, at Bemidji, Minnesota, serving as the prototypical giant statuary.[26] Certainly the do-it-yourselfers who built and managed the roadside services clad in exceptional buildings articulated no rationale for their buildings beyond surprising the passersby to snare their patronage. Observers after the fact have offered explanations of those curiosities. Three of these observers, Robert Venturi, Denise Scott Brown, and Steven Izenour, explained that those curiosities could be understood as signs communicating a worldview, not only a commodity or a brand. They categorized buildings as either a "decorated shed," a utilitarian form adapted outwardly to prevailing decorative tastes, or a "duck," a building shaped like something else. Diversity and egalitarianism was communicated. Instead of architect-designed buildings intended to convey the architect's message, buildings along the strip were informally designed and subject to whatever meaning the viewer took of them. Architecture became symbol. And of this postmodern aesthetic beginning in the 1960s, no buildings were more effective than those of the roadside strip, according to Venturi and his disciples.[27]

Ever larger architectural ensembles facilitated the development of strip buildings as signs. Shopping centers afforded opportunities for sign postings. New enterprises required identities; signs contributed. Supermarkets, one of the primary founders of the outlying shopping centers, led in sign innovation. Their signs gave commodity information but also gave visual directions permitting the companies to manage the flow of customers both inside and outside the store. A state-of-the-art setting in the early 1950s was the Stalcup supermarket in Kansas City, where painted bulletins and poster billboards on the parking lot circumference synchronized with sales signs inside. Lamps strung overhead guided the parked customers into the store. Numerous other stores located billboards throughout town and beside freeways, often several miles from the store itself, in hopes of bringing customers to their destination.[28]

Out front of the stores, signs were built taller, larger, and with uncommon shapes. By the 1950s, scientists on contract to the sign industry rendered elaborate mathematical formulas about the nature of various signs and their visibility under varying circumstances. Speed in getting the billposted message to the driver and passengers naturally deserved any signman's careful attention. Cars at 20- to 35-mile-per-hour speeds in town afforded observers less time to read a sign than was true when traffic before the automobile traveled more slowly. Some advertisers in the 1950s, however, concluded that faster cars and multiplying demands on the driver's attention rendered the sign maker's task more urgent. "Speed in Conveying Message Becomes More Vital Each Year" the editors of *Signs of the Times* headlined one article in a three-part series on scientific sign making. Signs became increasingly less a vernacular art of the untutored. At the same time, corporate identities induced visual uniformity from strip to strip (fig. 24).[29]

Taking Venturi's admonition to learn from Las Vegas about how the strip functions as a sign system, Alan Hess has most effectively turned attention to how its buildings and what in common parlance is known as signs worked together to symbolize. Hess elucidated that the unconventional, electrical spectaculars and parking lots gave new meaning to Las Vegas's common elements. Themed architecture on buildings set well back behind the big signs and parking lots hailing automobilists to turn in laid claims to the exotic, unusual, or preposterous. Various cultural subjects of long standing — the frontier West and knighthood, to identify but two — bait the traveler. Its signs crackling with vitality, Las Vegas was initially a brainchild of vernacular design genius, and the apotheosis of the building as sign on the Vegas Strip was a key component in foretelling a new urban paradigm.[30] Richard Francaviglia observed that the strategy of rendering architecture as signage for the

FIGURE 24
Holiday Inn made itself virtually synonymous with roadside lodging in the 1950s with this sign sought by travelers, as shown here in Liberal, Kansas, in 1983.

product or service sold within advanced by the 1960s onto small town Main Streets as well as rural stretches. But by the 1970s conformity with surrounding buildings returned to design in many small towns.[31] Whatever its future extent, the array of buildings, signs, and parking lots created a visual ensemble exciting and inviting. They did not amount to chaos, as traditionalists decried.

Notwithstanding technical and symbolic accomplishments, the outdoor advertising industry never achieved self-certainty throughout the century. A seventy-five year retrospective in *Signs of the Times*, marveled at "the industry's knack to *stay* in business."[32] At various intervals during the last thirty years, the sign industry believed that magazines and newspapers eroded advertising sign accounts.[33] In the 1950s, television curiously caused no alarm when the new medium drew considerable advertising. Unprecedented increases in sign revenues throughout the decade failed, however, to allay fears that renewed efforts to impose roadside zoning to control sign location brought threats against private property rights and national economic vitality.[34]

In the 1960s, highway beautification (which we discuss further in chapter 10) confronted the sign industry along the interstate system, renewing the tendency for sign regulation. The industry compensated for postings lost along the interstates with postings elsewhere. James and Karen Claus started their Institute of Signage Research in the wake of the highway beautification lobby, and it was so effective in arguments for the use of signs in a free economy that the institute became a legislative branch of the National Electric Sign Association in 1979.[35] Thereafter, the thesis that signs constitute the best kept secret in advertising was articulated more clearly than before.[36] The sign industry, meanwhile, continuously took advantage of its opportunities. Outdoor advertising earned considerable income in the 1980s from tobacco accounts shifted from television and radio to avoid a federal ban on tobacco advertising in those media.[37] Like a musical refrain on the century, during the sluggish economy in late 2000, billboards and street and transit signs again proved their unique saliency. Billboards especially were seen by more people than the print media and television because longer commuting time gave people the chance to see the signs, while print and television circulation declined.[38]

In the latter half of the twentieth century, sign company ownership took several different directions common to business in general at the time. Some long-established names disappeared. In 1963 the General Outdoor Advertising Company sold out to Gamble-Skogmo, Inc., which owned stores selling auto supplies, sporting goods, electrical appliances, radios, light hardware, paints, and clothing, but only a year later, the advertising sign plants were sold to various buyers. In 1985 the Patrick Media Group bought Foster and Kleiser, in keeping with a trend toward companies doing business in several media. Outdoor advertising's effectiveness was improved as a consequence of synchronization with the other media. Sensitivity to more localized markets and varying lengths of showings, instead of the common thirty-day period, came with synchronization. In 1979 Gannett Company, which owned newspapers and television and radio stations, bought Combined Communications and established the subsidiary Gannett Outdoor Advertising. In 1996 Outdoor Systems, Inc. (begun 1980), purchased the Gannett sign subsidiary and, nearing the end of the twentieth century, was the nation's largest "out-of-home" media company in North America. At the end of 1997, it operated about 231,000 bulletin, poster, mall, and transit advertising displays. Lamar Advertising Company, an exclusively outdoor advertiser begun in 1908, ranked third simultaneously but jumped to first place by 1999 with 130,000 billboards and 90,000 highway logo signs.[39]

New materials and techniques persistently took up much of the sign industry's resources. New materials and new techniques, however, were not always implemented as soon as they were available. A lag of several years often passed before many were convinced of an innovation's usefulness, and then a rush to apply it often followed. Neon, for example, waited until 1928, a decade after it was patented abroad, to be applied throughout the United States, where roadside applications became especially popular. Lighting, plastics, and light-reflective material, all introduced in the late 1930s, also experienced a delayed takeoff period. Practical considerations outweighed any urge to insert a new product into marketing for the purpose of rekindling Depression-ridden profits. It cost money to convert existing plants to new techniques and materials, and the industry had to be educated about their uses. Fluorescent lighting, light-reflective material, and plastics were not vigorously employed until the emergence of the postwar consumer economy in the late 1940s.[40]

In prosperous times, some techniques were quickly adopted and quickly abandoned or became but a minor mode. In the prosperous 1950s, fluorescent and phosphorescent paint illuminated with ultraviolet or mercury lighting was used, but only between 1950 and 1952. A modified technique, "blue light," extended the basic idea of the original "black light" into the present, but it is not commonly used. Day-glo graphics appeared in the 1960s, only to fall from popularity thereafter. Fads and fashions clearly stimulated the industry as well as sober calculations.[41]

Different display boards were cheaper to introduce and use than new technologies and satisfied the advertising imperative for constant novelty at a more superficial level. Cantilever construction, for instance, enabled advertising to appear at sites where conventional billboards could not stand. Cantilevered boards waited until after World War II and the introduction of portable cranes to mount them. They have been used widely since then. Raymond Loewy, having achieved fame in the 1930s as a founder of the new field of industrial design, successfully introduced a newly designed panel for posters in 1946. A gray and green molding, at first with a gold line inside the gray and a white mat around the poster, won instant approval from the Outdoor Advertising Association of America and became standard for many years. Cutouts reappeared in the 1950s, Hood Tire's campaign of 1917–1918 representing a rare early use. Stronger structures and more durable materials facilitated their reintroduction.[42]

Numerous sign techniques and materials poured into the market over the last thirty years. Some older ones have regained popularity, most notably neon and custom-painted signs. The former waned in the

1950s and reemerged in the 1990s. The latter, having lost ground to the types produced in quantity, began to recover in the early 1970s. Flexible-face fabric, which is durable and adaptable to numerous locations and sign sizes, has been slow to be adopted, remaining a material that only big companies have preferred since its introduction in the 1970s. Demonstrated utility and cost-effectiveness again were not the only criteria for adopting promising new techniques. Overall, some innovations helped excite visions of a stronger industry. Backlit posters beginning in the 1960s and computer designs have been foremost among these. Many more new techniques and materials could be listed, but this chapter is not intended to catalogue them all.

The cumulative effects of those introductions, both within the outdoor advertising industry and on society, however, were essential. Amidst the exuberant adoption of these innovations, so numerous and diverse as to almost bewilder insiders trying to make sense of the industry's recent history, tensions emerged. Among them were new questions. Could sign making recover its craft orientation? Were specialists in the outdoor medium more effective with it than general managers acquiring it to complement a broad range of other media in order to attract business from prospective advertisers? Old questions persisted. How could outdoor advertising companies convince advertisers that the outdoor medium was a cost-effective bargain? What limits did challengers of aggressive business ethics have in trying to restrain an industry so obviously serviceable?[43]

■ Signs have performed uniquely for all highway travelers. In giving information, they foretold what lay ahead, be it at a temporary stop or by alluding to what could be obtained after the journey ended. Expectation stirred. Travel became more than a destination. Signs did that without providing tangible essentials on a road trip, namely, fuel, lodging, and food, and signs did that without the total sensory experience that roadside architecture provides inside and outside. Signs are material culture, but they are the most abstract aspects of the roadside in acting upon travelers. They distill reality into an idea whose fulfillment is deferred. It became a logical consequence, when other circumstances enabled, to perceive landscapes as signs, that is, for the semiological triangle to be developed to unpack the roadside's meanings. Although that triangle provides an analytical scheme, it is heuristic. It centers the understanding of signs in the individual analyst's thinking. It follows that signs are freighted with subtle, subjective, and contingent understandings. Signs, whether bills posted or landscapes, are very much matters of *internal* concern belying their external dazzle that people often find sufficient reason for them.

Although signs function synchronically and abstractly, they evolved diachronically and so forcefully because they are tangible and act upon the primary human sense, vision. When the material culture of signs was transplanted from city streets into the open countryside, the landscape was made anew but looked only slightly different than it had before. "Tackers," "daubers," and the like were jarring in contrast with the earthen tones of dirt roads and weathered farm buildings. Billboards, the key sign determinant of the progressively different highway landscape, however, overpowered viewers by contrast. They were bigger and less easily ignored than smaller signs had been. Signmen sold billposting services on the basis of the billboards' potential for conveying impressions because passersby could not avoid them. Billboards are foremost in modernity's proud visual culture.

The automobile roadside with billboards in place plays to a host of other modern values. It helps separate the image in public view from the work "behind the scenes" that produces the subject of the image. Much of modern life becomes an artful facade to entertain the affluent, who can divert their attention from labor and the hard-working or have-nots. Could anything be more literal in this regard than the billboard? Its supporting structure is best when unseen or prettified in lattice aprons or artful frames. When effective, billboards could be passed at high speeds and yet deliver memorable impressions. Communications consequently became a realm of serious study and crafty contrivance. It was no longer permissible for signmen to do the best they could; they were expected to be unerring in maximizing the benefits of outdoor advertising. Plans grew more comprehensive as multimedia companies bought the stalwart founders of the outdoor advertising industry by the end of the twentieth century with the intent to meld outdoor advertising in campaigns utilizing a range of media.

The preference for art and distaste for blight set parameters for what competing businesses resolved for the collective look of the roadside. Histories of the billboard have paid far more attention to the best designers and the changing fashions of their craft than to the business owners' practical workings in response to critics of billboard blight. But leading entrepreneurs conducted an unromantic and strenuous twin strategy, to persuade their own ranks that billboard content and location should be discriminating as well as to persuade advertisers that billposting was *the best* advertising.

This meant campaigns of well-honed efficiency bridging tensions between change and stasis. Self-regulation to ward off attacks on billboards accused of blight dictated expensive gardening around billboards remodeled periodically in search of the best looking ensemble. Selling the outdoor medium necessitated procurement of the latest circulation data and incessant searches in new technologies for signs of improved durability, flexibility, and popular appeal. No formula for the best arrangement of all those factors could be achieved. Novelty accrued, and signmen also valorized it. Anything looking the same for long, it was thought, must surely be passé and useless. Displays changed regularly, lasting a standard thirty days early in the century and for shorter durations later in the century. Combine this reality of advertising contracts with the constantly innovated technologies, and it is apparent why signs, by their agency alone, made of the roadside a frenetically changeful landscape. Despite the constant showing of traditional themes in new guise and commodities of old vintage, viewers, like the motoring fans of Burma Shave signs, often waited for a changed display. Signs carried the highway's restlessness onto the roadside before the services there, ensuring both the delights of novelty and the dissatisfaction with it. If Burma Shave fans looked for new messages, they longed for Burma Shave signs in any case. Constantly changing outdoor displays rivaled the countervailing quest for permanency of place. The exchange also helped erode faith in planning based on long experience or history, because much of what was routinely shown in outdoor advertising were satisfactions either of a future or nostalgic perfection waiting to be purchased.

In the next three chapters we address the many other sign uses besides outdoor advertising for business. Those other applications left substantial, though sometimes transitory, marks.

PART TWO SIGNING PUBLIC PLACES

4 : Traffic Signs

Traffic signs, the subject of this chapter, are among the most commonly seen signs along the automotive highway. Their dour reminders to drivers and pleasant service to seekers of direction suffice, raising little doubt about the information, their necessity, or their reliability. Perhaps one finding one's way through strange territory may cleave to a sign in looking for guidance. Mostly though, traffic signs are taken for granted. But why do they look as they do? Who put them there? In short, what is the history of those key contributors to roadside geography? Gordon Sessions answered many of the questions about the "firsts" of traffic devices, and his book stands as the premier history of the subject up to the time of its publication in 1971.[1] The following offers instead a broad outline of answers rather than an encyclopedic explanation in the manner of Sessions's fine compendium.

Government, of course, at its various levels in the American system of federalized authority has the exclusive power to direct traffic on the public right-of-way. Moored to their classical liberal heritage, Americans have viewed their government's role minimally and deliberately divided power between numerous centers — federal, state, and local — to fend off usurpation. Thus, government generally does what other authorities cannot do or are unwilling to do. Directing traffic did not evolve from the premise that government was unfairly invading the sovereign sphere of private life but instead that people had a right to expect travel to be safe and easy.[2] No issue about the roadside has come closer to achieving absolute unanimity than the issue of who controls its traffic signs. Engineers, whose aura of objective and universalizing knowledge Bruce Seely has shown to be a major determinant in highway construction, and roadside beautifiers, the group most subjective in its standards about the highway, tacitly concurred: government must set the standards, post, and maintain traffic signs.[3]

These basic functions seemed reasonably conferred on government because the consequences were so essential; in a word, automobile driving was a complex task with life-threatening consequences. A team of traffic sign engineers in 1965 appreciated that "most people are so familiar with driving that it is assumed to be a simple human performance. On the contrary, when analyzed into a number of tasks which are carried on simultaneously and the functions which the human is performing, automobile driving proves to be a most complex human performance."[4]

Not the automobile by itself, but the various transportation forms — pedestrian, horseback, horse-drawn, streetcar, and bicycle — mixed with automobiles in America's big cities to produce traffic chaos by the start of the twentieth century. Sometimes lacking rules, facing indifferent enforcement, and exacerbated by narrow streets, the various transportation forms and people on foot competed for space with occasionally deadly but almost always congesting results. Traffic stalled; people and horses were killed or injured. Signs became but one factor in a series to reform traffic. William Phelps Eno, who is commonly regarded as the father of uniform traffic rules for having published an article on the subject in 1900, for example, saw boarding and unboarding carriages at theaters and keeping all vehicles to the right-hand side of the street of equal importance with a number of other problems. However, the increasingly dominant automobile and truck consequently drew most of Eno's attention as a leader in traffic regulation.[5]

At no time did people fail to appreciate the destructive potential to property and the lethal potential to life that the automobile represented, if gone awry. Manufacturers seeking profit through mass production were aware that their success depended on the conception of a dependable machine within wide financial reach and comfort with it at large. James Flink explained how easily Americans adopted the automobile compared to the self-propelled trolley car and bicycle that came before, partly by pointing out how little resentment the auto faced. The very contrast, however favorable for cars, overshadowed but did not expunge some degree of public reticence about the car. Previous experience with trolleys helped urbanites overcome earlier fears of machines on city streets. Automotive vehicles nonetheless contributed to a general concern about street traffic jeopardizing people and property.[6] In a sense, all of society was investing in the automobile even if, at first, only a few were actually buying models.

As both automobile ownership and automobile accidents mounted, reason grew to look for ways of accommodating this new mechanical servant more peaceably. The National Council for Industrial Safety, the brainchild of a group of concerned industrial leaders and government agencies founded in 1912 to diminish injuries and deaths to railroad and factory workers, broadened its mission in 1914 to address the rising toll from traffic accidents and dropped the restrictive modifier "industrial" from its name. A formidable power in the nascent profession of traffic engineering, the American Association of State Highway

Officials (AASHO), founded in 1914 to influence federal highway legislation, also soon turned toward highway safety. With motor vehicle deaths nearing twenty thousand annually by 1924, the National Safety Council and AASHO combined with the U.S. Chamber of Commerce, the U.S. Board of Standards, and the American Engineering Standards Committee to have Secretary of Commerce Herbert Hoover convene the National Conference on Street and Highway Safety.[7] Obviously, when traffic accidents were seen to threaten the nation's economic vitality in the twenties, one of the nation's most robust eras of business and business culture, steps were taken to curb the automobile's life- and property-threatening capabilities.

Public education became the watchword. Reporting accidents became a staple in the news media (fig. 25). Private enterprise joined the recurring campaigns to stop "the slaughter on the highway." Self-interested businesses like insurance companies advertised in print to persuade policy makers and drivers alike, the former to have signs posted literally on threatening roads and the latter to heed them.[8] *Signs of the Times* and public officials collaborated in efforts to persuade outdoor advertisers that the medium for which they had contrived strategies so effective in profit-making along highways could be applied with commensurate benefits in education about driving safety. The National Safety Council, too, was convinced that "safety could be sold," and *Signs of the Times* argued for the latest incarnation of the consumer culture, with billboards for safety "to obtain happy living and the benefits that the highways could bring."[9] Thus, the urgency of traffic safety became a leitmotif of everyday life, less dramatic perhaps than the shifting national policies of foreign affairs and domestic legislation, yet needing drivers' and pedestrians' constant vigilance and pushing officials to find new ways of fulfilling people's right to safety on streets and roads.

Safety has not been the only reason for traffic signs. Signs were also responsible for guiding the unfamiliar in increasingly complex and distant places. Urban expansion beginning in the nineteenth century brought with it longer reaches across diverse neighborhoods for the daily rounds of work and the journey to work. Gone was the comparatively simple access on foot to the requirements of life and recreation. A designer of urban street lighting in 1901, before the profusion of cars and trucks on city streets, understood that "the persons to be informed are not only strangers to the locality, but passengers in a swiftly-moving trolley car, who know quite well the street they are traversing, but who wish to know the street they are approaching or passing." The need seemed obvious for street-name signs easily read in the daylight

FIGURE 25
The collision of old and new technology, the railroad and the automobile, on unmarked country crossings became a special concern. The Buffalo Motorist, *for example, tried to alert readers to the rising death rate at crossings. That was twice as high for the year ending June 30, 1916, as the year before. Source:* Buffalo Motorist *10 (Nov. 1917), p. 27.*

and illuminated by night. Unlike countries where the government might expect the owner of a street corner building to display a street sign, American municipal authorities stepped into the void with municipal signs.[10]

Like their urban forerunners, roadside signs lagged in design and construction until shortly after highway improvement. "Good roads" interests crystallized among bicyclists in the 1880s and gained further support in the 1890s from the Grange and advocates of Rural Free (mail) Delivery (begun in 1896). But it wasn't until after formation of the federal authority on road survey (in 1899) and a federal construction program (with the Federal Aid Road Act of 1916) that a uniform sign system seemed necessary. Legislation for the construction and protection of "guideboards and guideposts" was introduced for Rural Free Delivery roads in 1919.[11] The Federal Highway Act of 1921 formulated a practical method of national highway finance. Uniformity of highway appearance would result finally from the act's financial incentive to grant aid on the condition that the recipient accept the proverbial strings tied to federal standards.

Signs became uniform nationwide as the result of a line of reasoning. Starting as a means to a set of ends, standardization eventually banished the idiosyncrasies of locale that characterized much of preautomotive America's roadside. The chief of safety engineering in the National Bureau of Standards during the watershed between the earlier era of largely statewide sign policies and the later era of largely nationally uniform sign policies illustrated how standardizers simply slipped into tyrannizing the sense of local place: "If vehicle operators are to be expected to comport themselves in conformity with local regulations and practices, these regulations and customs should be substantially uniform in all municipalities."[12] According to this reasoning, local pathways could remain serviceable only if they succumbed to national prescriptions. In this pragmatic fashion, modernity's characteristic standardization found its way onto America's traffic signs. The actual sequence of events was more convoluted.

While public authorities moved toward a single system of traffic signs, privatism prevailed. In 1912 the Lincoln Highway Association, a model of success for the numerous nonfederal highways following, started work on a system of uniform markers for the first transcontinental highway for automobiles. If realized, the Lincoln Highway would constitute a stunning reversal of existing conditions. At its birth, the Lincoln Highway's approximately thirty-three hundred miles were an amalgam of city streets and country roads lacking common signage. Local boosters at various points throughout the nation formed their own private road organizations. Their highway construction and sign posting for the new automobile, truck, and bus traffic reached its zenith in the mid-1920s.

Although the work of private organizations preceded the government's, where the latter were effective, they demonstrated efficiencies that strongly suggested the future of well-marked roads lay with them. Their signs could answer the elementary question on every driver's mind — "Where am I?" — at a time when indifferent road surfaces inconvenienced travelers. Maryland led the way in the early 1920s — in order that "he who runs may read" — by posting signs with maps, mileage to major destinations in a hundred-mile radius, notification of dangerous points on the road, and driving recommendations for avoiding trouble. Thomas Cusack, an outdoor advertising leader, contracted to build and maintain those signs to Maryland standards. Maryland also posted lengthy descriptions of several "Rules of the Road" on the assumption that most drivers would willingly adhere to them so long as they knew them.[13]

Since clear and instant communication was the first order of business in highway signs, it followed logically that standardized signs were one way to achieve this. Standardization programmed drivers to habit, enabling their rapid reflex rather than preliminary puzzlement. Mass production weighed equally in the reasoning on behalf of standardization, for with standardization the costs to interested governments, customarily prey to attacks against too much taxation, could be minimized. The more costly one-of-a-kind sign crafted for each location became avoidable. Numerous manufacturers marketed look-alike signs beginning in the 1910s. *Signs of the Times* pushed its merchant readership toward the prospect of increasing sales by stocking ready-made traffic signs plus the usual commercial signs.[14]

Various solutions for the safe and efficient movement of traffic having been found and automotive consumption apparently a staple in America for a host of reasons by the 1920s, public officials quickly devised and administered a program for the signs proven most promising of traffic relief. Private highways privately marked seemed antique. Hoover's conference of 1924 concluded with a report upholding the saliency of signs in achieving highway safety. Signs, it was advised, would be most effective when they were uniform throughout the nation. The federal government should lead the way to national uniformity by adopting standards that were simple, had the least wording to make them easily understood, and depended largely on distinctive shapes, symbols, and colors. Conferees also devised a set of colors for the various traffic functions. The AASHO adopted the sign standards, and a uniform vehicle code and a model traffic ordinance also soon resulted from Hoover's conference. Some states, such as Ohio, already employed a similar sign system, but the American legacy of local authority still frustrated advocates of traffic sign uniformity nationwide. In 1925 the Joint Board on Interstate Highways, whose twenty-one members from state highway departments and three members from the federal Bureau of Roads were appointed by the Secretary of Agriculture, recommended adopting a code for numbering interstate highways and the Mississippi Valley Association of State Highway Departments' set of shapes for highway signs. Here emerged the template for street and highway signs throughout the rest of the century (fig. 26). The Joint Board never invited input from the private highway organizations, because it regarded them as obstacles to progress. The

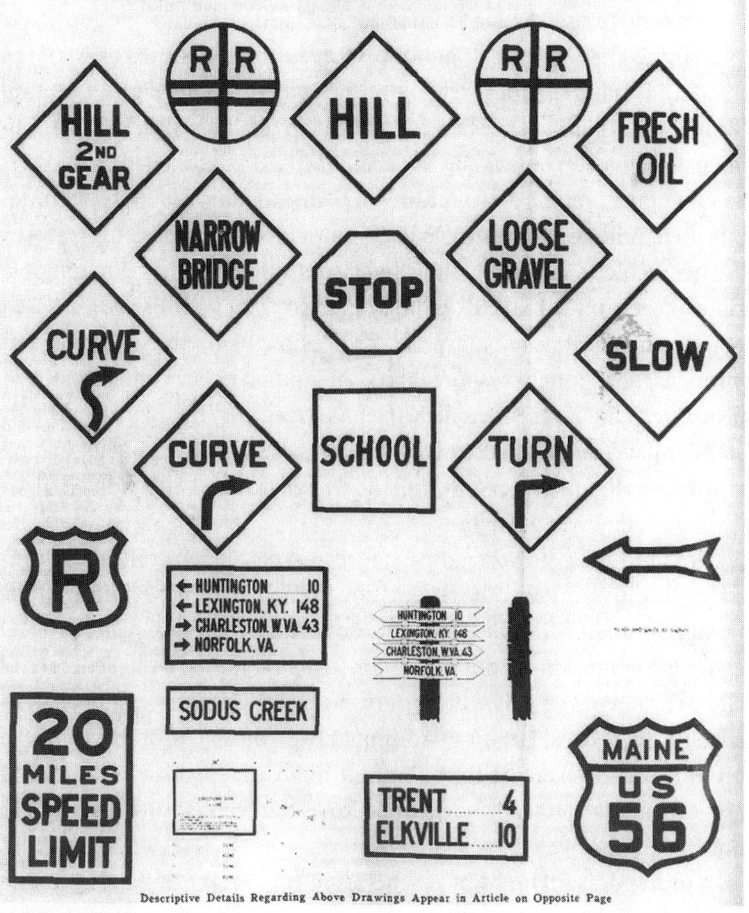

FIGURE 26

Shape and color reinforced each other to produce a readily distinguishable and easily learned multipart classification of signs: circles, diamonds, octagons, and squares for regulation and warning; and shields and rectangles for guidance. Source: "Uniform Highway Marks Now Become Reality," Signs of the Times *51 (Nov. 1925), p. 8.*

Lincoln Highway Association, behind the best publicized of all the named highway forerunners to the imminent national highway system, held out for retaining the names of the highways along with the numbers. Although its aim had shifted from private to public financing very soon after its formation because of the high cost and slow rate of private construction, the association sadly resigned itself to extinction, in

1927, when it had run out of options. In the pivotal year of 1926, the states rapidly adopted the current system of federally signed and numbered roads. The AASHO published a uniform sign manual the following year and refined it in 1931. In 1929 the National Conference on Street and Highway Safety, speaking for urban authority, adopted most of the AASHO's manual. Comprehensive agreement about signs for rural highways and urban streets was achieved in the *Manual on Uniform Traffic Control Devices for Streets and Highways (MUTCD)*, published in 1935.[15]

Out of the first third of a century in their experience with the automobile then, Americans had evolved a complex program facilitating travel by the new vehicles as conveniently and as untroubled as they thought possible. It certainly resulted from no one person's or group's preconceived grand scheme. Private enterprise, professionalism, bureaucratic commitment, government responsibility, and popular alarm collaborated in protracted and intermittent dialogue. As a result, by the 1930s nationally standardized signs shepherded travelers reliably to and safely along numbered alignments of government-designated and -maintained highways that ribboned the nation.

Not until 1944, however, did Congress, typically reluctant about central authority, empower the federal government to impose uniform traffic signals or the *MUTCD*. The requirement that a state adopt the *MUTCD* came as a condition for obtaining construction funds in the Federal Aid Highway Act of 1944. The body of highway officials and engineers that advised the Federal Highway Administration on the formulation of the *MUTCD* was divested of formal authority in 1979 as part of a general move against federal advisory committees but reconstituted itself in an unofficial capacity.[16]

A protagonist of highway aesthetics in 1932, at the beginning of one of the nation's great spurts of road building, measured civilization's progress in highway signs: "all signs along highways indicate, by their increasing or decreasing numbers and types, the trend of progress or lack of progress in modern highway development."[17] He felt that commercial signs signaled a slide backward for a civilized people; whereas, well-marked roads meant progress. Of whichever kind, highway signs symbolized more than their explicit messages. Commonsense demand for highway help, distinct from commercial advertising, had greatly multiplied the variety of signs throughout the nation. Transportation experts classified them into (1) regulations that imposed restrictions, (2) warnings that pointed out hazards, and (3) guides that gave information about a route's designation and directions to other destinations.[18] The latter could be rendered ancillary, so influential did the uniformly signed journey itself become.

REGULATIONS

> Terrible automatic red and green traffic lights, which bring ten thousand yards of street traffic to a stop at one time and allow more than a hundred cross streets to function, extend their dictatorship over the whole metropolitan area and are an affliction to one's nerves. I was irritated, depressed, made ill by them.[19]

Thus, disapprovingly did the master architect Le Corbusier react in 1946 to the vaunted standardization of twenty years before. The interstates, too, of more recent vintage, have come under similar fire.[20]

Firm regulation through uniformity was an unquestioned good at first. A safe and steady flow between pedestrians and automotive vehicles and between those vehicles in opposing directions initially presented the foremost challenge to those devising traffic regulations. Signs, especially semaphores and lights, were among the most effective early expedients, with whistles as adjuncts in some sign operations. Traffic towers were the first widely used remedy on the urban scene, their popularity lasting into the late 1920s. Traffic towers positioned at strategic locations near the automobile-pedestrian exchange sheltered police who operated stop-and-go signals from elevated sentry boxes overlooking the street. Introduced in Paris in 1912, the tower or "crow's nest" gained the greatest American publicity from its application in New York City, where it was introduced in 1919. Detroit had installed six towers between 1916 and 1919 using semaphore arms with "stop" and "go" on their faces, similar to those used on railroads, but New York City's proportions alone triggered more lasting memories.

New York also demonstrates how solutions evolved from trial and error. A survey disclosed that the city's seventeen busiest intersections were jammed with 255,842 vehicles in a twelve-hour period, far exceeding even London, England. Police posted at hand-operated semaphores along Fifth Avenue's traffic-congested one-and-one-half-mile heart (of its entire seven miles) had failed because, without coordination between the police, some stretches teamed with traffic while others were virtually empty. The city traffic commissioner, an unpaid official who temporarily left his medical practice out of personal interest to conduct the survey, directed construction of a tower with red, green, and orange electric lights eighteen feet above the midstream of the vehicles crossing Forty-second Street and Fifth Avenue. The orange light commanded movement on the street and a halt to cross streets. The green signal reversed the order on the streets and red warned that the signals were about to change. The system solved the problems and the city went on to install more than fifty towers before the precious space they occupied in the center of the traffic stream forced their removal beginning in 1929, as ever greater numbers poured into the city.[21]

If signals high overhead worked, why not try them low on the ground? Milwaukee installed a series of "mushroom type" lights in the pavement that combined the advantages of relative indestructibility and no damage to cars passing over them.[22] Those devices in New York and Milwaukee, however, represented the exceptions. It is generally conceded that Detroit introduced the first stop sign in 1915, having attempted since 1909 to have police regulate traffic with hand signals. The extraordinarily high incidence of horse-drawn vehicle accidents having provided the initial impetus in Detroit, it soon became the motor vehicles, for whose production that city became synonymous, that necessitated regulation of the traffic growing everywhere. Posts with globes atop were highly susceptible to destruction on the glutted streets, but the more durable corner-post signals became one of the most preferred forms. In 1921 Houston was the first to use a traffic light (semiautomatic) on a light pole. The next year, Houston employed a series of those structures electrically synchronized at adjacent intersections.[23] Overhead signs hanging between posts across the street from each other proved a useful alternative.

What drivers are most required to remember about stop-and-go signals is the color and meaning of the lights. A police officer in Salt Lake City is credited in 1912 with inventing, albeit manually operated, the first modern type of traffic signals. Five years later, an electrician working with him linked a succession of lights to coordinate traffic at a uniform speed. The red-green-yellow triad we have programmed our minds to obey since is not a symbolic system derived from nature. Highway traffic signals derived directly from railroad signals, where red meant stop because it contrasted best with adjacent signals and startled the eye most effectively in the hope of averting disaster. William Potts, a Detroit police officer on duty at the corner of Woodward Avenue and Fort Street, is commonly acknowledged as first applying the red-green-yellow combination to automobile traffic. It was his personal contribution to Detroit's apparently general exploration of solutions in the late 1910s for the new problems of automotive traffic. Most cities had adopted yellow with the now common red-green system when New York City conformed to it in 1924.[24]

When the American Engineering Standards Committee adopted the triad in the following year, it became the national standard.[25] The electrical lights that simultaneously became conventional required protection for use around the clock and to function without error. Lights were encased first in comparatively ornate boxes echoing the waning Victorian designs in machined goods and were reduced to spare utilitarian forms as the twentieth century unfolded. Visors shielding the light from glare and sunlight were consistent features throughout the various permutations, dictated by the utmost desire to make clear the intended signal. Lenses illuminated by available light could easily be mistaken for the wrong command, causing an accident.

Traffic engineers devised more than stoplights to reduce accidents. For street intersections, where an estimated 67 percent of the automobile fatalities occurred in 1934, one manufacturer sold signs that illuminated an arrow indicating the direction vehicles could go when the light turned green in addition to the red-green-yellow lenses shown above them.[26] "Through streets" permitting travelers access to residential neighborhoods were made safer for pedestrians by adding pavement signs with "stop" beside the standing stop signs.[27] "No passing" zones served a similar function on rural highways, but a problem remained because the states used three types of signs for this purpose by the late 1930s.[28] Manufacturers, responding to the hunger for ever greater safety, sold to car owners driver-operated devices that signaled stopping and turning. Mounted on the outside of the cab on the driver's side, they were synchronized with the driver's device inside the car. Municipal authorities could buy pedestrian-controlled devices to interrupt the regular sign cycle to permit street crossing.[29] Inventors continued creating various other improvements amounting to an entirely new traffic sign technology by century's end.

WARNING

Warning signs went through much the same development process as did regulation signs. Idiosyncrasy within the states and between states characterized warning signs until the late 1920s. Before then one could find warning signs posted by governments and private organizations alike. Representative of the way effective marking programs diffused, the Highway Sign Committee of the Association of County Highway Engineers meeting in Trenton, New Jersey, in 1918 recommended that states adopt neighboring Connecticut's danger signs. Some states also adopted the International Road Congress's catalog of pictograms for warning drivers. But, as one highway engineer noted only two years before the uniform sign system reigned nationally in 1925, local volunteer associations such as the named highway groups were remarkably dependable for posting good warning signs. The Automobile Club of Southern California, a private club of sixteen thousand motorists as of 1919, maintained one of the most exemplary of the early systems for traffic signs, including those for warning (fig. 27). Within five years, on the eve of Secretary of Commerce Hoover's national meeting on highway safety, Colorado and Utah had granted southern California's private auto club permission to post 3,500 of its variety of traffic signs inside the adjacent states.[30]

Notwithstanding the adoption of federal standards in the late 1920s, local peculiarities persisted due to a combination of budgets and local attitudes about change. Reflector buttons of various types existed for the earliest traffic signs, but a particular brand of reflector button introduced in 1931, for example, combined durability, low cost, and tremendous reflectivity — showing 75 percent of the original light from the oncoming vehicle.[31] It appears to have achieved instant acceptance among public officials, while lagging in commercial use.[32] Another manufacturer was simultaneously successful in selling "highway lighthouses" to highway departments. Highway lighthouses featured a lighted beacon with a warning sign atop a large pedestal on which private advertisements could be posted.[33] Civic-minded advertisers were called upon to provide these "sentinels of safety" without charge to the highway departments. Some states, however, prohibited such public-private combinations. In 1940 Washington, D.C., introduced a celebrated system coordinating traffic signals and "Walk! Don't Walk!" signals for pedestrians at busy Thomas Circle. Regulation signs like those minimized the use of warning signs.[34] As highways yielded more effortless driving, the need arose to be sure drivers paid attention and did not succumb to "highway hypnosis." North Carolina, often a leader in highway innovation, introduced a set of warning signs unconventional in that they deviated from the AASHO sign standards. Instead of diamond-shaped signs with the warning "Danger — Curve," North Carolina in 1956 experimented with a series of signs using slogans and rhymes. If early highways were too unpredictable, it was learned that a little bit of variety could be a good thing in a highly rationalized landscape.[35]

The perfect warning sign system driving engineering vision eluded grasp, in part because of the changing relationship between numerous

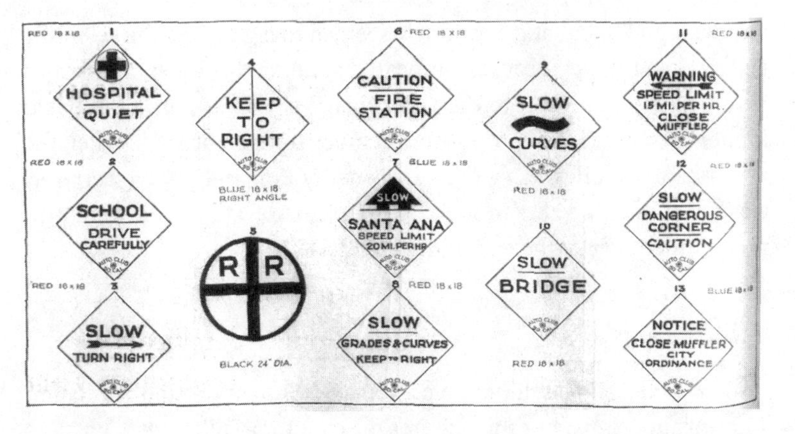

FIGURE 27

Blue lettering on white enamel was intended for clarity and endurance on the Automobile Club of Southern California's warning signs. Source: "How California Marks Its Highways — 60,000 Direction Signs Placed," Signs of the Times *31 (Mar. 1919), p. 44.*

factors. While early automobiling, for example, was almost entirely a daylight activity, improved road surfaces and automobiles increased driving by night. Scientists undertook elaborate experiments to make night driving safer. Their findings advised numerous small changes in the dimensions of lettering, sign distance from the road's edge, tilt of the sign toward oncoming traffic, and the relationship to other road signs. Pavement signs with letters of exaggerated height in proportion to narrow width became a reality defying the common sense of most sign makers. The heady beginnings of the interstate system in 1956 encouraged renewed discussion of using symbols instead of words, as many believed symbols to be more rapidly understood than words. The symbols were not adopted, however, because others countered that the larger the number of symbols, the less likely readers were to remember them. Electronic signs with computer-changed warnings about accident blockages, fog, spilled loads, and slippery pavements were introduced in 1965 by the California Department of Transportation on the Santa Monica Freeway between Los Angeles and Santa Monica. *Signs of the Times* reported in 1974 that the devices should be considered for congested expressways everywhere.[36]

Warning signs thus improved driving conditions, but traffic engineers remained as diligent in the search for better results as the drivers were advised to be in watching for and acting upon the latest signs. State-of-the art safety improvements were not always possible to institute, even when they were perfected, since sufficient funds for their purchase and installation routinely seemed to lag behind the scientific curve. Nonetheless, progress appeared genuine, for horse-and-buggy fatalities, especially due to railroad crossings, were estimated to be far higher than for cars in proportion to miles traveled by 1975, when the rate for automobiles was 3.6 fatalities per million miles. That certainly was lower than the 28.2 in 1920 on the eve of the standardization program undertaken at the end of the decade.[37]

GUIDANCE

Directional signs seldom saved lives and property, as did regulation and warning signs, but this did not mean automobilists were any less demanding of them. On urban streets poor lighting made good street signs even more essential.[38] The automobile brought similar problems to the open countryside. In 1918 Wisconsin pioneered a system of state route numbers marked with a *W*, symbolic of the system, inside an inverted triangle. Those signs were placed on utility poles, culvert shoulders, trees, or rocks. Wisconsin added a "Highway Information System" by the early 1920s.[39] Accordingly, commercial clubs, hotels, garages, and automobile clubs could obtain highway condition reports weekly for a seasonal (June to September) fee of ten dollars as an adjunct to the numbered road system.[40] Wisconsin's chief highway engineer called the five-thousand-mile system a "kindly reminder" for travelers.[41] Michigan next established its own numbered routes on painted boards, followed by Illinois with boards and aluminum signs (fig. 28), and Pennsylvania and Missouri with cast iron signs. North Carolina simultaneously started a simpler plan of marking roads radiating from each county seat according to letters "A" through "D" for the ordinal compass points.[42] Ohio, it was conceded, however, erected durable signs for the most fully marked system of highways by 1923.[43] New England states set the pace for a tricolor sign system throughout the entire region. Red designated east-west routes, blue the north-south routes, and yellow the diagonal or secondary routes. New York was on the verge of joining the system when Connecticut's entry won *Signs of the Times*'s attention in 1919. Ten-inch bands of color over fourteen-inch-wide bands of white on telegraph and other poles at one-mile intervals "clearly direct traffic along the main route," it was reported.[44]

Local boosters assumed authority for marking where state programs lagged behind before the federal system was implemented. San Antonians, for example, posted roadside signs publicizing both the city's tourist attractions and maps of its streets. Named highway organiza-

FIGURE 28
Illinois customarily displayed guidance signs on telephone poles, as shown in this mid-1920s view in the state capital, Springfield.

tions provided an interim solution by marking their alignments, but often provided the least direct route because nearby towns and villages lobbied successfully for meandering through their Main Streets. Then too, many named highway organizations were too poor to consistently mark their routes. Whereas the Lincoln Highway Association exemplified the high end of privately marked automobile trails, the Yellowstone Trail, running from South Dakota to Washington, was both marked and blazed in many locations by its founder working alone. A chasm divided the leading from the laggard states, for, in 1925, by the time the interstate highway board adopted the standard sign code, some of the states most insistent on it — New York, New Jersey, Michigan, Wisconsin, California, and all of New England — allowed no private road markers.[45]

Following the adoption of the uniform route markers and mileage destination signs, further improvements turned from sign format to materials. As vehicles traveled farther, new reflective materials were called upon to direct light farther for earlier notice to the driver and with less glare.[46] Budget-conscious highway bureaucracies found reasons to introduce lightweight, strong, yet economical overlaid plywood signs. Oregon, for example, saved 20 percent by using plywood rather than other materials, and one supplier reminded highway departments that, with plywood signs two-thirds the weight of others, their frequent relocation to keep current with changing alignments made them very economical.[47]

The interstate system came with new technologies for the many guidance signs required in a culture of increased diversions and sprawling settlement patterns. Franklin Steel Company of Franklin, Pennsylvania, perfected bolted-base steel channel signposts in cooperation with the Texas Transportation Institute at Texas A&M. Those structures avoided the complications of installation and hazards to workmen that the previous U-post presented. The New Jersey Department of Transportation pioneered the testing of "breakaway" signs in 1970. Simultaneously, engineers tested and introduced for common use the long-span sign structure. Arms as long as two hundred feet overarched multiple-lane traffic. Bolted behind the lanes' bordering guardrails, long-span structures satisfied higher safety standards at the same time as new aesthetic standards. The AASHO initiated tests on the structures in 1967 and pronounced them satisfactory. In the early 1990s the Interstate Highway Sign Company of Little Rock, Arkansas, manufactured slightly more than half the green-and-white directional giants overhanging the nation's interstates. Life expectancies of seven to twelve years for such signs and the minute variations required locally ensured a brisk trade in the signs travelers came to expect. Lowercase lettering, long believed to be most easily read on traffic signs and bestowing a high-quality appearance, was adopted for the interstate system and restricted to destination names.[48]

Recreational and historic site signs became more important complements of the guidance sign category as the twentieth century ended. Vermont, Oregon, Minnesota, and Nebraska reported major tourist programs requiring new highway signs in the 1990s. The hazard to unfamiliar motorists and the fourteen to sixteen dollar per square foot cost for each installed sign motivated a field test for the most effective sign at Devil's Tower and South Pass City, Wyoming, beginning in 1988. Results were provisional, the engineers recognizing the complexity of the task. More research would be required to understand repetitive signing, to determine the best sign location between one and ten miles in advance of the site, and to establish a criterion for those sites meriting "advanced" or guidance signs. National Park Service highway signs were upgraded in the early 1990s from white letters on a wood board to recycled plastic and fiberglass signs decorated with artistic logos and requiring little maintenance. Although the park service allocated only 3 percent of its budget for signs manufactured commercially (as opposed to those produced by the federal prison industry), signmen eagerly anticipated a growing demand, both from the park service and the states.[49] Signs facilitated big public business.

Custodians of the new business of tourism, thus, were virtually unanimous about the imperative of traffic signs locked into a highway system. Seeking individuality for the sites where they offered recreation, their managers ironically introduced the old efficiencies of uniformity to get people to come. A periodic *Car and Driver* magazine contest on photographs of eccentric traffic signs was humorous counterpoint to the dull sameness of the signed highways' machined efficiencies.[50]

■ Highway travel regulated and guided from roadside signs is a subject interlacing several themes familiar by now: commercialism, standardization, and residual government. Rising out of urban places beset by congested traffic, puzzled city governments willingly conferred public authority on ingenious individuals trying independently across the nation to solve the twin problems of traffic safety and property loss. Ad hoc solutions or ones adapted from Europe to local circumstances amounted to numerous field-proven alternatives by the 1910s. Highways then confronted hesitant state authorities with prospectively long traverses and their consequently huge budgetary demands. Businesspeople and boosters filled the void at first with named highways marked at the expense of privately raised funds. Private groups achieved little more than auto trails, and governments were pressured for an end to mounting life and property losses. The combination pushed commercial interests out of road building and roadside marking in favor of federal and state governments. Key states introduced their efficacious regulation and guidance signs with the result that standardization took hold of traffic signs throughout the nation. The pioneering look went out of the road as place.

Complaints about the monotony of trouble-free interstate driving are a luxury of a presentist people without living memory of past frustrations in travel. Not only highway surfaces but signs reduce the anxieties of travel to a minimum, the whole of the roadway and shoulder merged into a fluid flight path. People just go, little wondering why it is so easy to find their way and arrive so safely. Out of disparate localities, some state, some regional (New England and California), an unquestioned assumption of transcontinental nationhood was forged. Federal aid road and interstate shields reaffirm the unity at every intersection, cloverleaf, and mile marker. Highway and street markers heighten both the automobilist's and the pedestrian's conceptualization of place. Functional to be sure, traffic signs contributed substantially to formulating a place of seemingly alienating uniformity.

We now turn from the unloved but utilitarian traffic signs of national community to the types of communities for which Americans yearned spiritually. We investigate how signs nurtured this yearning.

5 : Signs and Community

In this chapter we address how signs reflect and instill a sense of belonging to a community. In a nation committed to the welfare of individual citizens, the importance of the community as an ideal is easily dismissed. Certainly contemporary Americans, it would seem, value their personal prerogatives, defining success in life, for example, largely in personal terms. Yet, begrudgingly perhaps, even the most self-centered of individuals admits to being part of a larger social collective: the nation as a whole, one's fellow residents in a city, town, or neighborhood, one's work associates, or one's fellow church congregants. But just what is community? How has it been symbolized in landscape? What is it that signs contribute to its recognition? Community is a concept that almost any American will agree to understanding. It is a word that is so widely used that careful definition at the beginning is essential.

Sociologists early wrestled with the community concept, because they considered it fundamental to understanding human socialization. Ferdinand Tönnies's book *Gemeinschaft und Gesellschaft*, published over a century ago (1887), is widely accepted as the first rigorous contribution to the self-conscious dialogue about the term *community*. Through the succeeding twentieth century, sociologists in America turned to less historical and philosophical questions in favor of empirically grounded studies, leaving community to isolated scholars in other disciplines, writers, and laypersons. Geographer Donald Meinig, for example, found in the popularity of Thornton Wilder's play *Our Town* proof that Americans, at least in the 1930s, were widely drawn to inquiries into the community concept.[1]

Global depression in the 1930s rekindled anxieties about an economy dedicated to private gain, turning voters overwhelmingly toward the New Deal, with policies like social security rather than free exchange in the marketplace. The turn toward community was in truth more than a passing political reaction of one decade, for it centered in the very circumstances of modern consciousness, what cultural theorist Siegfried Kracauer termed the "transcendental homelessness."[2] There it functioned in dialectical tension with the quest for meaning in lives loosened from the medieval strictures of a status-based society and exacerbated uncannily in America with its strongly espoused individual freedoms in a contract-based society. Individualism and communitarianism structures a basic axis of American culture.[3] Americans valued a classless society and avoided an established church, with the result that each person was to find his or her own way in the world. Democracy promised not only the advantage of privately made gains safeguarded for private use but also the uncertainties of separation from others in gaining and securing those rewards. In the increasingly industrialized and urbanized society of the nineteenth century, many Americans hungered for security. They compensated by pursuing similarities and continuities in other individuals and in rituals performed with them. The dialectic endures.

A decade ago, political scientist Robert Booth Fowler, having reviewed the long and complicated dialogue about community, afforded us a definition at once reliable and succinct. Community, in Fowler's thought, assumes a group of people "sharing in common, being and experiencing together." It also implies that "an affective or emotion dimension" bonds the community.[4] Reason plays its role too, for in reflecting on why one participates in a particular community, one will conceptualize reasons. Community does not abandon reason to the corporate or governmental bureaucracies, where roles are itemized and lines of authority delineated, but community ties depend more fundamentally on "a deeply felt experience" than do those other forms of association.[5] Thus, we have, to elaborate in our words, a community of interest. We further add the corollary that a community of interest makes physical boundaries of occasional importance, but not of necessary importance, because affective cohesion does not always function with physical place in mind.[6] Some communities are defined in their territorial extent, but all communities are linked by mutual interests wherever their members are located.

This constructed nature of community can render it hard for outsiders to identify, because they often grasp for tangible traces like clothing, food customs, and signs to seek out otherness. In their seeming vagueness, Fowler's elements — commonality and emotional above rational satisfaction — may leave some scholars hungering for more precision, but the virtues of Fowler's terms are twofold. Not only are his conclusions grounded in serious historiographical analysis of the community concept, but they also correspond to the customary response to intellectual conundrums. Ask the average person "What is community?" and she or he may be unable to put it in words. Yet community is something that average persons believe they know when they see it and either enjoy or shy away from when they experience it in practice. Many will rest comfortably with Fowler's definition of community.

Our definition implies through whose lens we wish to look for how signs of community work in the landscape. Our own cannot help but come first, of course, but we discipline it in quest of the signs most people see or make in the landscape. We are concerned with representativity, the average, the commonplace. Those who have engendered the resurgence of thought about community, intellectuals contending in the political realm since the 1970s about the loss of community, provide only uncertain guidance. They do accept the historical challenge of locating community's origins. In community they almost unanimously prescribe the remedial alternative to classical liberalism's inherently destructive marketplace values. Although these present-minded critics disagree about the exact symptoms of the malaise, they believe Americans en masse have generally valued self-aggrandizement above commitments to any joint social, religious, or even economic undertakings. They find the state substituting its terms for the codes of propriety, responsibility, and welfare that community should properly uphold.[7] Robert Putnam's best-seller *Bowling Alone* (2000) renewed the old jeremiad that community life had declined, although he departed from received wisdom in his display of graphics and arithmetic to prove his point.[8] Clearly, community — its history, current vigor, and its very validity as a concept — continues to engage a lot of serious reflection, debate, and writing.

The differences of opinion between liberalism's contemporary critics about the occurrences of community in the past, however, can raise doubts altogether about community's viability. Depending on which authority you read, community has lapsed since the creation of the Constitution or since the populist movement at the turn of the nineteenth century or has never been more than a minor voice.[9] On the contrary, some recent procommunity theorists find in de Tocqueville the father of the belief that Americans have long struggled with the consequences of community versus individualism and found in the privacy of family and religion realms to satisfy their strong appetite for shared meaning and purposes above self.[10] Self-interest and consumerism do not guide Americans as a nation of individualists, asserts Robert Wuthnow, who sees the recent growth of the small-group movement as a manifestation of Americans' enduring commitment to community.[11]

The enduring individualist theme in America may be demonstrated by the tendency among many Americans to consume styles especially suited to the unique personality advertisers insist we all embody or by the defense of one's civil liberties from the encroachments of government authority. At the same time, the material culture of ordinary signs in the landscape offers proof that various commitments to community do thrive. These commitments perhaps show more sporadically and often more privately than expressions of individualism, but they show just as certainly and with the same involuntary vigor as individualism. Community's several names command instant allegiance. What few Americans do not proclaim faith in one or more of the following institutions: "God," "country," "family," and "neighborhood"? Listen to orations on ritual holidays, public dedications, or preachments on holy days. You will hear the quaternary faith repeated and see reverent listeners silently nod their agreement.

This chapter is organized around the ways and extent to which signs participate in this four-part social matrix: nation, neighborhood, religion, and family. Because we are concerned with how signs function regarding those four institutions, not with those institutions themselves, we address them in the order of those making the most use of signs and proceed to those which make less use of signs.

NATION

In a nation without a common ancestry or belief in one and founded by negotiators at a constitutional convention to establish a government principally ensuring that contractual relationships are upheld legally, a degree of conditionality has colored national life. No dimension of community in America has made grander claims on geographical extent and tended to suppress doubt of it more than the advocates of nationhood. Perhaps for those reasons, patriots most formidably display their signs of community. None of the other four institutions of community in this chapter has been more dramatically challenged in fact than was the United States's very existence in the Civil War, when a strong rival fought to dismember the country. The flag of the Confederate States of America still rivets many observers nearly a century and a half after the Civil War ended. Although the Confederate nation-state no longer survives for partisans to swear allegiance to, the "Stars and Bars" bitterly divides. Some despise it for symbolizing racism, while some honor it for pride in local authority. Nations have always been synonymous with territorial integrity, and no more effective symbol of internal cohesion, popular vitality, and separation from others across boundaries has been asserted than making, waving, and posting the national flag. At every perceived challenge to their shared government and way of life, Americans have unfurled "Old Glory" in all its forms—flown, draped, in bunting, painted, or on adhesive stickers. The reaction can be traced in its formalities to at least 1877, when Flag Day was first celebrated, then to 1943, when Congress adopted a detailed code of flag etiquette, and to 1949, when a national Flag Day was enacted. Wilbur Zelinsky observed that flag display became a mania in the 1890s.[12]

During the first decades of the twentieth century, when immigrant laborers importing radical ideals appeared to threaten the capitalist status quo, challenged elites called for flag displays. They wanted to tutor the immigrants, and they believed that a sign's nonverbal power is great, that its meaning requires no explication. Gazing upon this talisman can alone suffice to work its emotive power, and persuade it surely is meant to do. Tutorship implies the typically hierarchical relationship as well: immigrants, though they be adults, and children alike have to be taught patriotism. Whatever affinities with the nation are stimulated in viewing the flag, they are constructs of social intercourse and do not lie dormant waiting to be aroused. The flag is part of a sign system reinforced wherever another flag is seen. Society leaves casual conversation and formal civics lessons to teach what the *feelings* of patriotism refer to. Traditionalists often lament the supposed decline of patriotism in peacetime among bystanders who no longer feel an emotion upon sight of the flag, as in a parade. They question not only the viewer's patriotism but implicitly question the flag's capacity to teach by repetition in this sign system. Some have argued that flags should be used sparingly, reserved for public institutions, and that the flag's code of etiquette should be rigidly honored.[13] When flags are displayed spontaneously, in war time or during events of national stress, those who display the flag not only declare their commitment to the country but reaffirm their hope that they can construct a similar commitment in those viewing the flag. Teaching patriotism becomes a shared responsibility fulfilled entirely in brandishing this nonverbal symbol (fig. 29).

Abstracted representations of the national flag have served as effectively as the flag itself. The "Lone Star" flag of Texas, with fewer stripes, just one star, yet all three colors of the American flag, was successfully promoted for a national symbol surrogate at the end of the media-saturated 1990s. One roadside salesman of soap dishes and bar stools with flag motifs explained that they reflected national virtues including loyalty, strength, and bravery.[14] Material advantage has customarily blended with patriotism in the vernacular flag display distinct from

FIGURE 29
These young muralists in Dana, Indiana, avowed their patriotism and bonds to a transcontinental nation tangible only in their mind's eye. They painted their flag a few days after the terrorist attacks in far distant New York City and Arlington, Virginia, on September 11, 2001.

formal etiquette, perhaps because profit-making has been the means to materialize the "American Dream." In the early stages of the Great Depression, when it was mistaken for a "recent stock market deflation," the newly born Outdoor Advertising Association of America posted fifty thousand panels in 17,500 cities with the "Forward America" display.[15] Six years later, in 1936, in the depths of the acknowledged and catastrophic depression, the association allied with the National Association of Manufacturers to sponsor a series of posters endorsing "the world's greatest opportunity for economic independence" as the "American way" and entwined home ownership with the nuclear family and enjoyment of highway travel in the family's own automobile (fig. 30).[16] Just as Americans benefited from their economy, they were obligated to make it successful by rushing out to buy. Material goods defined the nation in an essential manner. For all the public service displays that companies anonymously made, some could not resist declaring their identity on however small a portion of the sign.

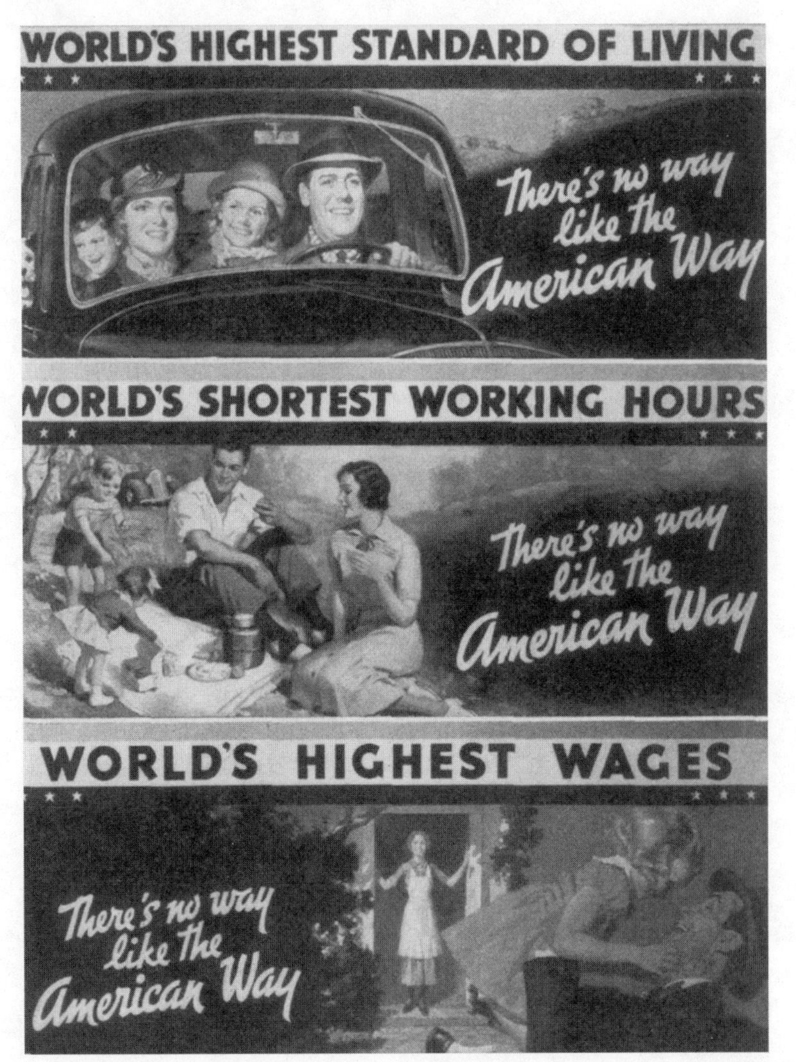

FIGURE 30
These posters by Foster and Kleiser have become popular culture icons since discharging their originally inspirational purposes. Source: "O.A.A. Votes to Back Manufacturers in Campaign of Industrial Leaders," Signs of the Times *84 (1936), p. 9.*

TOWN AND CITY

We take liberties with the common meaning of *neighborhood* to explain how signs function in that realm of community between the very public expression of nationhood on the one hand and the very private testimonies to religion and family on the other. We expand the spatial dimension of neighborhood, which customarily encompasses a section of a town or city, to include an entire city. This is not far-fetched when you consider the ways in which city boosters conceived and displayed signs to outsiders in the automobile age. Boosters tried to depict their economic sphere of interest in signs inviting people to enter a friendly place of close-knit fellowship.

Boosters were those frenetic individuals, almost always businesspeople, who, consistent with classical liberalism, believed their town or city of operation would benefit most from a positive reputation abroad drawing ever more business investment and travelers. *Boosting* referred to their ceaseless schemes, founded in the faith that continuous economic growth was the highest test of community virtue, that consequently a community should be boosted, or lifted, beyond financial subsistence.

Their mantra: "Boost! Don't Knock!" Their credo insisted that businesses should be mutually supportive, not destructively competitive. From aggressive self-promotion, overall benefit to fellow businesspeople and local citizenry alike would flow. Sociologist Claude Fischer put it pointedly: "In America, the free pursuit of the private good *is* the public good."[17] The same conviction manifested itself in the early adjudication of urban billboard ordinances, when lawyers and judges both believed society's best interests above all lay in protecting entrepreneurial rights. At this philosophical point, boosterism actually threatened community. Privitism has usually resulted in personal gains and eyewash about thoughts of the public good.[18] Critical self-examination bordered on defeatism. Anyone questioning commercialism could only be a detractor that peer pressure should stifle if not silence outright. Everyone in a city or town was expected to get behind its booster campaign. In an individualistic culture, the strategy was believed the simplest and quickest technique for arousing the sleeping citizenry accountable for the listless economy. Boosterism focused energies and sought to monopolize civic-mindedness.

Boosterism has not been limited to commercially defined communities seeking expansion. By the late nineteenth century, the development of the nation's rail and river transportation network had sorted cities and towns into an urban hierarchy, with each center having its own specialties. The coming of the automobile and its highway system galvanized a new round of booster vision throughout almost every city and town on or near an actual or would-be line of automotive traffic. Established tourist destinations reinvented in the new era of automobile travel stressed signs. The signs' importance could easily be gauged in the sign war erupting from Kentucky's bitterly factionalized politics.

When the Louisville Automobile Club secretary directed tourists away from Bardstown, state highway commissioner Ben Johnson's hometown, "Boss Ben" wielded state highway funds for political advantage and ordered state workers to remove about four hundred of the club's highway signs.[19]

Alerting the sign industry to a trend, the *Poster* magazine in 1928 observed that *Nation's Business* "recently declared that the most strenuous competition in America today was not between individuals or between corporations, but between cities for development and population growth."[20] Advertising in magazines and on billboards, California and Florida especially set out to produce economic booms, according to the *Poster*'s columnist, and in the process postured like underdeveloped economies or colonies isolated from the brusque mainstream yet offering amenities in paradise. Carl Fisher, a multimillionaire by thirty-seven, coinventor of an automobile headlight, and moving force behind the transcontinental Lincoln Highway from its inception in 1912, was simply the foremost Florida promoter in the state's economy when he developed Miami Beach. He started the Dixie Highway to facilitate the passage of travelers and potential settlers from Michigan to Florida, and especially Miami, and avidly endorsed the rash of highway and real estate subdivision signs posted to lure and direct migrants on the way.[21]

Meanwhile, on the West Coast, San Diego's businesspeople launched the first extensive and integrated advertising campaign for a city in the automobile era. Los Angeles took a step in this direction during 1914 and 1915, when the Automobile Club of Southern California, headquartered in Los Angeles, posted highway markers with distance and direction along the three thousand miles to Kansas City to facilitate travel and hoping as well to induce permanent settlement.[22] But San Diego's campaign advertised its virtues, not just its mileage and direction from a given sign. Following the city's penchant for grand boosterish schemes, as demonstrated in the nationally publicized Panama-California Exposition of 1915–1916 with its Spanish colonial city in Balboa Park, the San Diego–California Club formed in 1919, not only to boost the city out of the post–World War I depression but to boost it to the forefront of West Coast cities bidding for the tourist trade. San Diego's club proceeded routinely at first, placing commonly formatted advertisements in widely circulated newspapers and magazines, but began in 1926–1927 to build large and attractively designed billboards between Phoenix, Arizona, and the city to lure arrivals along U.S. 80.[23] An agent traveled the road and distributed maps to motorists and statements about current road conditions to facilitate travel at a time when road systems were still uncertain. Witnessing the slow descent of most hotels' physical plant and the reputation of their clientele, San Diego's missionary travel agent went instead to the new auto camps, where he identified, in the *Poster* columnist's words, the "better class of tourists" according to their cars and engaged them in personal discussions about San Diego.[24] Face-to-face communication, an essential ingredient in Tönnies's pioneering definition of community, reinforced maps and travel aids as signs. "Welcome into our community's life," if only for the trip's duration, they seemed to say. Hospitality of this type suggested temporary citizenship.

The coming of automotive transportation did not stimulate the discovery of this commercially based psychology of community that signs were intended to construct. Indeed, old understandings were cleverly adapted to the special conditions of the new transportation mode. Boosters inherited a well-articulated strategy. While the automotive highway system was still in its formative engineering and financial stages, boosters shrewdly realized how signs could help overcome the potential discouragements to travel through terra incognita. How could the traveler know beforehand where or if basic services were available in a distant location? How could the traveler feel assured that local merchants would deal honestly? Constructing an aura of community through signs meant to boosters better chances to profit from travelers than would be possible if travelers were left to find their own way to local merchants. *Signs of the Times*'s editors forthrightly declared in 1915, "There is nothing more welcome to the automobile or other traveler than the friendly road sign, particularly is it to the automobilist."[25] Nothing, it was calculated, would initially integrate the new arrivals into a caring environment better than a sign that declared where they were and where they were entering. "It rather puts upon him the obligation to patronize the merchants of that town if he is in need of anything they can supply." The circumspect traveler "*feels* that the people of the town are glad that he has arrived, that they appreciate his presence, that they will give him a square deal" [authors' italics]. To familiarize outsiders with local practices, such as speed limits, boosters in Marietta, Ohio, recommended signs saying "Please Slow Down." To clinch the sale of constructed and temporary community, Marietta's head of the Chamber of Commerce counseled that his counterparts elsewhere should post courteous exit signs: "Thank You! Come Again."[26] Happy visitors also made good ambassadors for a place once they returned home to informally advise people in their community of where to travel.

Although this sense of place so deliberately contrived contradicted the casual rapport in which sociologists and writers located a true community, boosters were persuaded of its genuineness.

Could communities along highways expand into long linear arrays of mutually interested population centers yet remain spatially distant from one another as automobility increased? The first new highways had stirred economic imagination, but their boosters stayed satisfied with the restorative powers within existing communities. What was the full capacity of highway signs for stimulating connectivity between communities once aroused?

Enthusiasts of the Route 66 and Lincoln Highway revivals beginning in the 1980s and 1990s demonstrated different rates of success in answering this question. In Pennsylvania, where the Lincoln Highway revivalists were most active, plans progressed for reeducating travelers exactly where the highway passed. Lincoln markers of uniform design, the fifth such effort in the highway's history, were promised by the Pennsylvania highway department in 1997.[27] The Lincoln Highway Corridor, a six-county region in central Pennsylvania, was declared in 1995 to preserve resources, both scenic and historic, and help organize future development, both commercial and preservation.[28] Road rallies brought faithful enthusiasts together annually to drive the marked highway in a shared experience.[29] Signs were imperative to this intended community of extended territoriality.

Route 66 witnessed a similar program for rekindling a sense of community but it achieved more dramatic effects than the other celebrity highway for two reasons. First, its mythical and lyrical traditions — it was widely believed to be *the* portal to personal fulfillment, as the Joads anticipated in *The Grapes of Wrath* and as foretold in the song "Get Your Kicks on Route 66" — seized a lot of average people and the media alike. Second, it inspired a succession of bards who further ennobled it with their own versions of its essentially nurturing community. Tom Teague, for example, closed his diary of a pilgrimage, *Searching for 66* (1991), thus: "Route 66 was more than a million yards of concrete, a path from Point A to Point B — it was a 2,400-mile long community. And in searching for that community the past four months, I had become part of it."[30] Modernity's pathos of estrangement ended, at least for this one person, and the highway performed its healing mission. Finding sections of the original highway for those seeking their own communion with the storied Route 66 became an essential and benevolent mission. Marking the highway with historic simulations of the federal "U.S. 66" marker along the original alignment became a popular government activity, with magazine writers regularly taking their cues from the signs to write short human-interest columns.

Highway logos reducing the identities of whole towns and cities to sign displays on the interstate routes demonstrate the continual reliance on those media for chasing the elusive essence of community many believe lost. Only time will reveal whether this goal is an illusion, whether signs move people beyond portent.

More coolly analytical than the highway bards, J. B. Jackson wrote in one of his last works that Americans crowded onto streets and roads, not in desperate acts of anomie, but because, in fact, that was where people had obviously thrived in harmonious community since the late 1940s. Urban streets most dramatically riveted his attention, perhaps because their bustle was concentrated so much more than on highways: "The neighborhood, such as it is, comes to life, and you begin to think this is a world where community and cars belong together, like bread and butter or ham and eggs."[31] Signed streets and byways were not to be taken for granted.

Signs became in those circumstances more than sources of information (fig. 31). For travelers seeking out those neighborhoods of self-

FIGURE 31

Boston's Dudley Street Neighborhood Initiative "Unity through Diversity Mural" in Roxbury, shown here in 1998, stated its ambition for community.

declared interest and for denizens alike, signs were proud emblems asserting spiritual rebirth. The National Trust for Historic Preservation's magazine, *Preservation*, periodically chronicled examples hovering in doubt across the nation.[32] Find community signs displayed and you find a healthy state of affairs.

Boosters had long before perfected several effective techniques implementing signs to attract business homeward. Humor was one. An insurance company in Detroit experimented in the 1920s with various kinds of posted information outside their offices to win new clients. Statistics about how many people were killed or injured did not seem effective. Passersby looked and moved on. An accidentally misspelled word on a sign, however, resulted in a number of letters to the company. Some declared the belief that it had been done purposefully, to provoke comment because "dry statistics" did not work. Others lectured the company about the correct spelling, believing officials did not know it. Because of these letters, company officials decided to try humor. Humor did increase circulation past the sign, if not business.[33]

Beauty also enticed. Businesspeople hearing Lorado Taft, sculptor and self-assigned evangelist for civic beauty, address the Recreation Congress in Atlantic City in 1922 could take heart that their advertising was not entirely selfish. For Taft, beauty and art had moral dimensions. Good citizenship began with children learning to pick up rubbish, Taft preached, and he lauded the Poster Advertising Association for its artistic displays, especially in small towns, where he believed little else was available to elevate taste.[34]

Before utilizing images, boosters had implemented "slogan signs" in the 1910s. They fashioned short phrases made memorable with mental images, often based on a local landform, local product, or, lacking either, a simple rhyme: "Galveston — the Treasure Island of America" and "Durham — Renowned the World Around."[35] *Signs of the Times*'s editors fanned the desire for such signs in periodic lists of cities that had recently added them.[36] Nearly two hundred cities had electric slogan signs by 1917, when the demand appears to have declined.[37]

The famous hillside "Hollywood" sign in the California city of that name is perhaps the best documented case of how a sign for commercial exploitation can actually produce attributes of a genuine neighborhood. Its origins in hype, pluck, and entrepreneurial relish were quintessentially Angeleno of the 1920s. In 1923 the transplanted midwesterner and 26-year-old ad man and promoter John D. Roche, Jr., happened onto the idea of a sign spectacular to sell real estate in Beachwood Canyon for the 640-acre Hollywoodland subdivision.

Roche had casually written "Hollywoodland" on the side of Mt. Lee overlooking the subdivision in a sales brochure showing homesites and streets for the subdivision. Harry Chandler, one of the subdivision's developers and publisher of the *Los Angeles Times*, saw the brochure, especially liked the name written on the side of Mt. Lee, and asked if Roche could actually build a sign there visible throughout all Los Angeles. Without any engineering study, Roche proceeded to identify the point of maximum visibility, had telephone poles brought by mules up Mt. Lee for each letter's piling, supervised between fifty and one hundred laborers, who chipped and dug the pilings' holes in solid rock and soil, built a special road to haul up the metal letters, supervised their fastening, and in sixty days completed the Herculean task for $21,000.[38] The subdivision was bought up. Later, four thousand low-watt bulbs electrified the sign, and in 1945 the realty company donated it to the city. In 1949 the sign's last syllable was removed leaving it to stand as most people think of it, "Hollywood" in a display one block long and letters fifty feet high. It fell into neglect beginning in 1939, when its maintenance fund was exhausted and its commercial use had long before expired. Public opinion always rallied behind preservation remedies. In 1949, when the last syllable was removed, the initial H, casualty of a storm, was replaced. By 1973, when the sign badly needed repair and faced the city's demolition plans, interested citizens again collaborated to save it. The Hollywood Chamber of Commerce and radio station KABC started the "Save the Sign" campaign to raise $15,000. R. Leslie Kelley, a retired auto dealer and creator of the leading price guide in auto sales, chipped in $10,000 and started a trust fund for the sign's perpetual life. Clinching the 1973 rally, the Los Angeles Cultural Heritage Board conferred official respectability on the sign by designating it "Monument No. 111."[39] Waging a campaign of public opinion and finance to save a community sign had reinforced the community's actual bonds. Once taking literal directions to a future destination from the sign on the hill, many now learned to think of their common past through it.

Signs on the edge of town, both to notify those coming through a gateway or leaving on the opposite side, have not been exclusively exploitive. Zelinsky correctly asserted that signs conveying a sense of place at the gateway are more for its inhabitants than for outsiders.[40] Sports records after all are innumerable and less distinguishable as they multiply, except to locals who can take pride in individualizing their otherwise likely forgotten town and assume membership in an ennobled civic pedigree (fig. 32). Richard Francaviglia's patient survey of Texas's vast space reports that its towns often secured their belonging

FIGURE 32
Chisholm, Minnesota, doted attention on its corporate physical prowess in this cumulative athletic record displayed in 2000.

in it through entrance signs shaped like the state and sometimes marked with the Lone Star in addition to each towns' local attractions.[41] Barbara Weightman, observing to fellow geographers the sign's frequent role in creating a sense of place in midwestern towns, emphasized how many had their name painted on water towers (fig. 33).[42] Rotary, Kiwanis, and Lions club signs at myriad small-town gateways may fade in the eye of the casual observer but not in the mind's eye of the club faithful or of town dwellers for whom the club's benevolent programs figure prominently in town life.[43] Reinforcing nostalgic mementos of small town togetherness, "Thanks for Visiting" or "Come Again" signs usher visitors out at the same time as they induce a fond look over the shoulder. James Parsons lamented the lack of systematic study of hillside signs but felt assured they have become integral to local community identity since rival college students seem to have created most of them early in the twentieth century.[44]

Signs perform critically at town entrances as well as at exits, because they are the one element of those passages' material culture most widely understood. Especially where a town's future seemed problematic in recent times, usually a small rural center, landscape architects and designers, often commissioned by town elders seeking identity for their charge in the acutely self-conscious consumer culture, articulated the rationale. According to one design team, every town's "image experience" is divisible into "announcement," "entry," and "exit." The

FIGURE 33

High above the observer, water towers obtain a commanding vantage point, as shown here in Buhl, Minnesota, in 2000.

point of announcement is "one of the most exacting of all image experiences" because it requires multiple satisfactions in a brief instant. Entry marks the transition into; it is a specific point of separation. Exit is more exclusive, "a feeling more than a place," comprehensible as a transition out into the country.[45] An experience, more surely felt than analyzed, and stages of passage simultaneously exacting and elusive underscore any architecture's or design's fundamentally subjective liabilities, for they can be mistaken unless explicit declarations of intent—signs of announcement, entry, and exit—are posted. Writer Kent Ryden pointed out the inherently liminal distractions of crossing geographic borders. Happening upon a relict marker between Connecticut and Rhode Island on Route 101 one morning, Ryden's imagination was stimulated to reflect an entire day on the landscape about the rich interplay of the marker's meanings in place and time, geography and history, ending with hurried recourse to preserving his impressions in writing in a notebook just before sunset.[46] Crossing over any line distinguishing a "here" from a "there" may be one of life's simplest acts, yet it is complicated by its consequences. How soon we learn that things do not unerringly go as we intend. Imagination's capacity thus stimulated for seeing several paths diverging from any boundary, however unambiguously its creators have designed a "correct" way of crossing, makes it almost certain that sign makers will continue blazing trails.

RELIGION AND FAMILY

We only briefly outline the relationship between signs and the practices of religious faith and families, not because of their standing in American life, which indeed is robust, but because until very recent times their signs have not been much shown. Faith and family have always figured prominently in the nation's past, and, because people have little disagreed about their importance, their presence was sufficient to speak for itself.

Religious signs have also been minor expressions of material culture until lately, because the constitutional premise of church-state separation has shunted matters of faith into private expression. Inheritors of the Enlightenment principle that knowledge derives from human reason and persuaded that divinely revealed teaching had produced bloody religious war in their recent past, the Constitution's founders denied organized religion any overt role in national affairs. To be sure, many belonged to a religious denomination and believed in some definition of God, but they did not wish to impose it on others. Religion's proper place was in church and the home (where family reigned). Living in harmony with innumerable faiths precluded any thought of an established church. Symbols of the various Reformed denominations, Catholicism, Judaism, and other faiths of smaller numbers were integral to their places of worship and liturgies. No need arose for their use at large.

The extent to which religion retreated into privacy, giving ground to consumerism as an organizing social principle in contemporary America, is clear from the tendency to think of sales and not souls as fit for advertising.[47] Is it proper for faiths to advertise? For the most part, however, demonstrations of faith were usually quietly informational, boards on sanctuary grounds announcing coming events or identifying symbols of faith on burial grounds.

Over the last two decades, religion has broken with this demur tradition. While the faithful continued practicing their ways, greater differences of opinion arose between radically different worldviews within sects than between sects, and communities of interest united members of different sects disagreeing with their fellow coreligionists. Two sociologists explained:

> what *unites* the orthodox and the progressive *across* tradition and *divides* the orthodox and progressive *within* tradition are different formulations of moral authority. Whereas the orthodox side of the cultural divide is guided by conceptions of a transcendent source of moral authority, the progressive formulation grants that authority to what could be called "self-grounded rational discourse."[48]

That chasm fueled the debates about abortion, homosexuality, the definition of family, prayer in school, and evolution, to mention some of the most volatile issues.

Family has undergone a similar fundamental reorientation at approximately the same time. Through the 1950s, Americans clung to a normative definition named the "modern family." In practice Americans had organized family life variously during the nation's history but they predicated a hegemonic ideal: as Judith Stacey wrote, "an intact nuclear household unit composed of a male breadwinner, his full-time homemaker wife, and their dependent children — precisely the form of family life that many mistake for an ancient, essential, and now endangered institution."[49] General impressions drawn from random observation and statistical proofs that the ideal is less practiced, in fact, have generated the widespread fear that the family is in decline. Activists prophesy social chaos without the modern family and prescribe the remedies in urgent campaigns against drug dependency, teenage pregnancy, child abuse, and birth control.[50] The alarms may have grown all the more extreme because for numerous generations, beginning in the nineteenth century, people often substituted their seemingly diminished channels for public expression through closer family involvement.[51]

Those embattled questions spawn sign making and display. Demonstrators bear graphic representations like "Stop the Baby Killing" and "The Family that Prays Together Stays Together." By literally carrying their cause in symbolic representations on signs, the adherents struggle to render a solution favorable to them. Expect to see signs so long as the outcomes remain in doubt in this deeply serious tangent of community as religion and family.

■ One of the most enduring dialogues in American history, regarding the degree to which Americans are involved in community life, not surprisingly shows itself in signs of many varieties. Is patriotism alive and well? Will Americans sacrifice their precious selves for the collective good? Are there national ideals even beyond dedication to one's fellow citizens? The flag itself, symbol of the whole national experience, and variations on it displayed widely and exuberantly in both formal and informal occasions, answers yes. Fretting about viewers unmoved by paraded flags only testifies how zealously some exhort and passionately need the signs of nationhood. Surely the elaborate code of flag etiquette testifies to the high degree to which many Americans are committed to an object of material culture above individual people. Some super patriots would readily sacrifice those who violate the idealistic code.

Does neighborhood as a branch of community resonate in signs? Commercial signs using one's expected dedication to a hometown or the nation to encourage the purchase of products or services abound. Boosters have habitually experimented with and refined signs to satisfy those purposes for many years. Signs along paths of entry also aim to ease travelers' tensions about unfamiliar ground. Where to buy? Who deals in quality? Are they honest? Outsiders are meant to feel like adopted insiders. Seekers of community free from private gain have posted signs of unity within rejuvenating residential areas and along interstate highways, especially during the last quarter-century of the historic preservation movement.

The superabundance of signs declaring public purpose and their general scarcity for religious faith and family, until recent disagreement, suggest that, when something is in question between groups, signs appear. Perhaps in a diverse, contentious, and publicity-oriented culture, the absence of any signs constitutes a silent agreement that "no news is good news." Signs should not be overlooked in search of how Americans think of their commitment to others.

PART THREE **SIGNING PERSONAL SPACE**

6 : Territorial Markers and Signs of Personal Identity

In modern America, the liberated individual has come to the fore, especially as a consumer of goods and services available in an increasingly globalized marketplace. Who we are has become very much a matter of how we live, and that, in turn, has become substantially a function of what we buy. It is as consumers, rather than as producers, that most of us see ourselves fulfilling life's ambitions. Indeed, through widely shared lifestyles, traditional social differences, and thus much social inequality, are seen to be diminished. We may not look alike or speak alike or even act alike, but we can enjoy having the same things. Race, gender, religion, national origin (or ethnicity), and other traditional aspects of social being remain important in assigning personal identity, but today such distinctions take significance mainly as they influence access to education, networking into jobs, and, therefore, the ability to earn and to spend.

As individuals, Americans constantly reflect upon who they are, taking the measure of others in assessing themselves. Mostly we assess one another at a distance, relying, therefore, on social stereotyping. Who we are is communicated by the kind of car we drive, the sort of house we live in, the vacations we take, and all the other evidences of personal consumption. Place is always involved in such evaluation. Houses are valued according to kind of neighborhood. Vacations are valued according to locale. Of course, such social meanings are not just place contained, but place constructed.

Signs posted in landscape affect personal identity in at least two ways. First, they instruct as to personal need. Advertising signs, especially, seek to prompt consumption, to instill a sense of wanting. At the same time, they seek to persuade as to a specific remedy. Advertising challenges the individual to self-assessment and to self-fulfillment. Second, signs enable people to better define the place-implications of their lives. Signs are especially useful in establishing the boundaries of place and in defining place-appropriateness. They can carry clear territorial implications, helping, for example, to assert privacy or, conversely, to intensify public interaction with others. Posting signs, rather than merely reading signs, enables people to call attention to themselves. All signs serve to regulate behavior within social settings or situations. They encourage individuals to abide by established norms, but in so doing they also provide base measures for violating those norms.

ADVERTISING SIGNS

Advertising rose to prominence in modern America as a kind of discourse through and about commodities. We thrive by appropriating from our surroundings selected goods and services. At base are particular material needs. As modern consumers, of course, we purchase most things rather than making them ourselves. In the process, we often lose sight of just how and where things are made and who makes them. The relationship between people and nature is one of commodification, things being related to people through market availability.

Advertising is integral to the defining of need, and thus to the creation of a market. Advertisements not only describe goods and services, but they connect their possession to implied states of human satisfaction, integrating consumers within a rich and complex web of social implication.[1] Advertisements mediate between things and people, making the possession of things (or not possessing them) the essence of personal self-worth. Goods and services stand as social markers, enabling people to relate socially through varying degrees of possession; advertisements set up this social connection.[2]

Ads tell us what things look like and how they can be used. More importantly, they suggest how one can personally benefit from owning or using those things. Depicted facial expression (among other aspects of body language), ascribed words of utterance, and the picturing of place as surrounding context are often combined in suggesting personal satisfaction. We variously internalize these appearances in response to the emotions they tap. These impressions, stored away, serve to predispose future knowing and future behavior—remembered appearances helping us, quite literally, to make appearances. Impressions absorbed from ads become part of our "looking-glass" selves, influencing not only how we look at others, but how we think others look at us.[3]

Advertisements exist in various worlds simultaneously: the world of the sender (the client whose product or service is being promoted), the fictional world of the ad (as depicted within the ad), the fantasy world of the receiver (the musings of the potential customer), and the real world of the receiver. It is the aim of the sender to sell something via the fiction of a story told, a story made to resonate in the receiver's psyche as reinforced fantasy.[4] The spark of energy that connects these various worlds is always ideological. Advertisers adopt models that instruct us about others. These models stand insinuated in ads, necessitated by the social circumstances that they help perpetuate. By *ideology*, we mean the often hidden ideas that buttress and support a particular distribution of power in society. Advertising constructs idealized images of people and depicts idealized patterns of interaction between them, thus positioning people socially.[5]

Advertising may well be the "official art" of capitalist society. Certainly, it is an art with clear economic purpose. In defining commodities and encouraging their consumption, advertising supports production. It sustains continuing rounds of investment and profit taking by creating and directing market demand. In the abstract, ads "arrange, organize and steer 'meanings' into 'signs' that can be inscribed on products: always geared to transferring the value of one measuring system to another," in the words of Robert Goldman. In this way, he continues, advertising comprises a system of "commodity-signs."[6] Accordingly, a Mercedes automobile or a Rolex watch is made to stand not just as a functioning device, but as a symbol of affluent status. Over time Mercedes and Rolex advertising came to emphasize this symbolic aspect so successfully that the car and the wristwatch serve today more as "signifiers" than as things signified. Because images are readily deployed as signifiers of social relationship, modern advertising teaches us to consume not just a commodity, but its "sign" as social symbol as well.[7]

What seems to characterize the economies of developed nations is the vast array of products and services available to so many of their citizens. But something else equally as dramatic in scale often goes unrecognized. Little thought of are the multiple layers of images and symbols that surround consumers in relating commodities to personal happiness.[8] Modern consumer society operates in a complex field of symbolization. Images of social respectability, desirability, and opportunity circulate as consumption-based imagery. Perhaps the most powerful of all are the forcefully graphic iconic images, especially the visual representations that firmly associate commodities with feelings of well-being through allusion or analogy.[9]

Although the history of advertising art has been written largely through consideration of magazine ads and television commercials, it is probably with posters and other forms of outdoor advertising that the evolution of iconic imagery is most clearly seen. From the beginning, billboard ads had to be concise in their communication. Posters could only be briefly glimpsed while people moved along city streets or rural highways. Motorists had only a few seconds to notice and to comprehend.

Thus, wordiness was eliminated in favor of hard-hitting slogans and visual graphics. Before World War I, products tended to be pictured so as to detail their manner of construction, artistic design, quality, and price. Products, accompanied by an identifying trademark or logo, were often pictured standing alone, abstracted against blank backgrounds rather than placed in geographical contact. Between the two world wars, however, the emphasis shifted toward overt use of social symbolism, products being depicted more in use. People in ads were carefully articulated to conform with prevailing social stereotypes, especially those that spoke of social success. Equally important, elaborated backgrounds offered contextualization: products, people, and evidence of their satisfaction linked through allusion to landscape and place. After World War II, the personalization of ads increased with the glamorous, the romantic, and the sensual given fuller play.

Consumption came to be represented as spectacle: advertisements, billboard ads especially, presenting social "tableaux" to be emulated. Of course, advertisements did not reflect life in America accurately, but distorted or skewed it in ways persuasive of product purchase. The preferred images depicted America not as it was, but as it ought to be. Ads reflected public aspirations, mirroring popular fantasies rather than popular realities. Advertisers realized that consumers would rather identify with scenes of higher status than ponder reflections of actual lives. As Roland Marchand observed: "In response, they often sought to give products a 'class image' by placing them in what recent advertising jargon would call 'upscale' settings."[10] Additionally, social aspiration was closely aligned with change: a clear bias in favor of modernity. It was not the old symbols of social class and status that counted so much as the new lifestyle symbols, the having of new things.

Through repeated viewing, and in combination with advertising seen in other media, the billboard ad was intended more to predispose than to convince. It was intended to incline people toward purchases rather than clinch sales. Advertisements were also intended to encourage a customer's continued commitment to a product line. But billboard images, in framing the huckster's pitch, predisposed viewers not only to product purchase and brand loyalty, but to lifestyle choices as well. E. M. Fairley's poem "Poster People" appeared in 1920 in the *Poster*, a trade journal oriented to graphic artists. In part, it read:

> Say, often when I'm kinda broke
> I'd like to be

One of these fancy poster folk
That you can see
On any board from Maine to Frisco
Riding in Coles or Super-Sixes,
Eating hot flap-jacks that she mixes
From posted Hecker's Flour and Crisco.
Oh to be a poster man,
With a pretty poster wife!
Like a carefree poster clan,
With a suit of Style-Plus Clothes,
And my feet in Onyx Hose,
And a Chesterfield between my manly fingers;
In a world where skies are blue,
I'd be happy, wouldn't you? —
Where the frigid hand of winter never lingers.[11]

PLACE AND PERSONAL IDENTITY

"Poster men" enjoy one another's company in the La Palina Cigar ad pictured in figure 34. The Congress Cigar Company of Philadelphia launched in 1922 a nationwide posterboard campaign linked, of course, with magazine and newspaper advertising. The idea was to show its brand of cigars as central not just to male camaraderie, but especially to the camaraderie of successful men who had "made good." "Crowd impression" was woven into poster design, journalist Philip Chandler reported. Messages were not overt, but fully covert. "It seems to be a natural reaction of the American people to somewhat resent a too forceful statement," he wrote, "but if the same statement can be pictorially presented in such a way that the message arrives in a more subtle manner, its effect has twice the strength of the more direct statement."[12] Such advertising pushed selected tobacco companies steadily to the fore. In 1922 some 65 percent of the cigar business in the United States was controlled by one hundred of the nearly twelve thousand cigar makers. In 1925 it was 80 percent.[13]

Characteristic of the ad's place in time, the depiction of successful businessmen lighting up their cigars is only minimally contextualized (see fig. 34). Barely discernible as place or setting is the business conference table. Slightly fuller contextualization is provided by an abstracted "frame," suggestive of a commercial street, added by the magazine's editors. After 1920, posters and other outdoor advertisements

FIGURE 34

A La Palina Cigar poster. A brief play on words associates a company's product with society's elites. In the process, pretentiousness in overstatement is studiously avoided. The company's cigars, the ad says, are well-made for those who have done well in "making good." Source: Philip Chandler, "La Palina Cigars 'Made Good,'" Poster, *16 (July 1925), p. 7.*

increasingly emphasized sense of place. Relations between people were fused through a focus not just on products and users, so much as on products and users fully situated in some geographical context. Social meaning was instilled to the extent that the depicted place was emblematic of one or another way of life. Implicit was a shifting away from sole reliance on traditional class and status differences toward increased emphasis on personal self-fulfillment and private fantasy built around the having of new things.[14] Social meaning was carried by the built environment depicted with its "to whom it may concern" messages articulated as social situation; sense of place was used to locate viewers socially.[15]

Sense of self is defined and expressed not simply through one's relationships with people, but also by one's relationships with place, or, perhaps more accurately, by one's relationships with people in place. A primary function of the sense of self is integrative, in that it organizes and unifies a person's behavior and experience both across situations and in the more immediate response to specific situations.[16] Human personality, therefore, contains an environmental or geographical component: a predisposing toward or preference for certain kinds of places over others. Advertisers learned not only how to amplify recognizable and widely shared place meanings in their ads, but, more importantly, how to use such meaning to sell things. Place, as social situation, is a kind of process. It involves exploring the behavioral possibilities of settings: people assessing actual places first-hand and, more importantly (and more frequently), assessing imaginary places through advertising and other images.[17]

SOCIAL STEREOTYPING IN ADS

Stereotyping involves relatively rigid, overly simplified, and often biased conceptions of reality. Stereotypes enter into social relations in that they invite premature and often thoughtless assessments of others in lieu of fuller knowing. Negative stereotyping often serves the insidious purpose of enhancing the stereotyper and his or her own fully self-serving self-image. By depreciating others categorically, one bolsters oneself categorically. Of course, stereotypes can be positive as well as negative. However, when they are incorrect, and thus potentially damaging to others, they are rightly condemned as morally wrong. Even positive stereotyping, when strictly adhered to, carries negative implications since it leads to the confirming of only shallow expectations, real learning being thereby diminished.[18] Everyone stereotypes in taking preliminary "reads" on people and on places, but discerning inquiry pushes to deeper levels of understanding.

Billboard advertising has relied on a relatively narrow range of social stereotype. Not that advertisers in general deliberately sought to depreciate certain groups (although some certainly did), but rather, most ad makers were simply overcommitted to using social cliché to maximize customer responsiveness. The typical American consumer was conceived as being male, white, and affluent. Mimicking and ridiculing others in enhancing white male egos was deemed acceptable. Through the 1960s, women were pictured in ads in highly restricted roles. Groups of men might stand around in boardrooms smoking cigars, but women, more often than not, were depicted standing alone in laundry rooms trying out one or another detergent, or in kitchens using one or another "labor saving" device.

Gender stereotyping in advertising depicted scenes scripted to emphasize the differences between men and women; it presented those differences as natural and thus as inevitable.[19] The nuclear family was assumed to be American society's fundamental building block, and households were assumed to be divided between male and female spheres of responsibility. Men were the breadwinners and, accordingly, participated fully in "public life." Women were fully domestic, being charged with raising children, tending to household upkeep, and otherwise supporting their husbands. Men were producers and women were consumers. Indeed, women were important customers, being responsible for up to 80 percent of all household purchasing in the United States by the 1920s.[20] However, this did not stop advertisers from emphasizing masculine values.

Ad makers, the vast majority of them male, did not employ gender-based stereotyping blindly. Rather, they sought to anchor ad images in systematic market research. What, for example, predisposed American women in their buying habits? What inclined them to buy this rather than that? Women were even hired as market analysts to help answer such questions. Wilma McKenzie concluded:

> The following are . . . the Chief instincts of women: Sex love; Mother love; Love of home making; Vanity and love of personal adornment; Love of . . . style, modernity, prestige, [and] reputation; Hospitality; Sociability; Curiosity; Rivalry, Envy, Jealousy; Pride, Ostentation, Display; Exclusiveness, Social Ambition and Snobbery; Tenderness, sympathy and pity.[21]

At the bottom of her list, interestingly enough, was "Love of beauty."

"It seems that the out-of-door poster has been the suggesting friend of many women," wrote another analyst, Evelyn Rucker. For example, "Kate has copied her most successful salad recipe from a poster and she brags about it. . . . An alluring poster is responsible for Martha's purchase of a sport model roadster." Women, Rucker asserted, were greatly influenced by outdoor advertising, since women seemed to respond to suggestion more readily than did men, perhaps because of vanity. "Then there is romance — and women are romantic. A poster presenting a moonlight scene with a wistful young man gazing over the gleaming shoulder of a dainty girl may awaken memories or longings, but it sells cold cream or powder or whatever it advertises better than any salesman could hope to do."[22] Indeed, many posterboards of the 1920s asserted forcefully this very point (fig. 35). Most women, it was supposed, really had only one fundamental goal in life: to find and keep a man. In doing so, women had to constantly appraise themselves as to desirability, especially with regard to appearances.

Beginning in the 1960s, many American women rebelled against being judged by appearances, against being considered objects, espe-

FIGURE 35
Ivory Soap poster advertisements. Source: Poster *18 (Sept. 1927), p. 18.*

cially of the male gaze. Targeted particularly was the depiction of women in advertisements. As critic Lucy Komisar wrote,

> Madison Avenue woman is a combination sex object and wife-mother who achieves fulfillment by looking beautiful and alluring for boy friend and lover and cooking, cleaning, washing or polishing for her

> husband and family. She is not very bright; she is submissive and subservient to men; if she has a job it is probably that of a secretary or an airline hostess. What she does is not very important anyway since the chief interest in her life is the "male reward" advertisers dangle enticingly in front of her.[23]

If feminine stereotypes in ads emphasized such character traits as passivity, complacency, and even narcissism, then masculine stereotypes emphasized power, performance, and precision. Men were represented as independent. They were seen as savoring freedom and as being adventurous. They were seen to be competitive.[24] Of all the products brought to the American scene early in the twentieth century, automobiles, perhaps more than any other salable product, appealed to the male persona. Evelyn Rucker's fictional Martha may have bought a roadster, but cars were really a male domain. "Automobiles are this country's phallic power symbol," Lucy Komisar observed, "and cars are used to prove man's masculinity."[25] Women, therefore, made logical props in ads geared to selling cars to men and in those ads selling things for cars, such as gasoline and tires. Certainly, the female portrayed as temptress reinforced male receptiveness, if only as an attention getting device. But women depicted as sex objects also legitimized male assertiveness and, conversely, served to question the manhood of men who did not buy into such stereotyping. Why was it masculine to portray men washing cars in a driveway, but a sign of henpecking for them to be pictured washing dishes in a kitchen?[26]

Throughout most of the twentieth century, advertising served African Americans very poorly. Blacks were often crudely caricatured on signs designed as if to make the reification of race a fact, if not of nature, then of society. "Aunt Jemima," "Uncle Ben," and the "Gold Dust Twins" were used to promote specific corporate brands.[27] White characters, such as "Betty Crocker," were also introduced, but they tended not to stand in advertising as exaggerated fictions. Black Americans were almost always shown subservient to whites or pictured doing menial things. Such images transcended mere symbolization to assume iconic status, serving the public at large as constant reminders of black inferiority. With the rise of the Civil Rights Movement in the 1960s, the nation's advertisers, some grudgingly so, began to include more realistic, and thus more positive, images of blacks. But at first this more realistic depiction was primarily found in advertising targeted to black markets. When mainstream media were used, it was often to promote, so critics charged, unhealthy products such as fast foods, tobacco, and alcoholic beverages.[28]

LIFESTYLE AND ADVERTISING

Advertising came to adopt what historian Roland Marchand has called a "therapeutic role." By encouraging new products and services, especially those with entertainment or leisure time implications, advertising encouraged Americans to identify more with lifestyle differences than with traditional notions of social segmentation. "Individual Americans," Marchand wrote, "never need feel themselves diminished or alienated, whatever the scale of life. They could enjoy modern artifact and style without losing the reassuring emotional bonds of the village community." Advertising emphasized a shared "culture of daily life" communicated through the constant barrage of advertising's "short-take" messages.[29] A steady diet of concise, carefully pitched advertising transformed, in another observer's words, "both the outer landscape in which we live and the inner one . . . by which we try to make sense of what is going on."[30]

Thus, the consumption of images, as well as the consumption of things, became the very essence of modern life. As geographer Don Mitchell observed: "It's not the flavor or thirst-quenching abilities of Coca-Cola that matter, it is the image we project while drinking it. The 'use value' of Levis is not that they keep us clothed, but the image they project (of youth, rebellion, and so forth)."[31] Today, consumerism is America's (if not the developed world's) prevailing ideology. It is not socialism (or any other "ism" or "ology") that has prevailed to capture the world's imagination so fundamentally.[32]

Consuming follows diverse lifestyle orientations. Pioneer market analyst Arnold Mitchell suggested nine lifestyle clusters defined around age, education, and income correlates. The bottom two-thirds of the American population he identified as "Needs-Driven" shoppers, who were subdivided into "Survivors" and slightly more affluent "Outer-Directed" shoppers (subdivided, in turn, into "Belongers," "Emulators," and "Achievers"). In the top third were "Inner-Directed" shoppers consisting, for example, of "I am Mes" (mainly young rebels from affluent families), slightly older "Experientials," and the "Socially Conscious."[33] Marketers needed to know that "Belongers" might buy station wagons or vans, but that "Experimentals" would more likely buy sporty cars. Advertisers also needed to realize, however, that different lifestyle groups sometimes bought the same things, but for different reasons. A consumer's lifestyle was likely to change with changes in age, education, and income. Shared lifestyle did not drive strict conformity in the marketplace. Nevertheless, through the very buying of goods and services, lifestyle preferences were constantly renegotiated. It was the advertiser's challenge to mold that negotiation as a form of symbolic interaction (fig. 36).

Billboards carry social messages scripted through poster design. A sign's success is not just a function of its graphics and its wording, however, but also of the quality of its positioning in the built environment. Location matters. Signs take meaning from the geographical context of landscape. In part, they are interpreted according to where they are seen. So also does landscape take meaning from signs.

What we would emphasize here, however, is the way advertising images, in their spatial spread, foster "geographies" of social comprehension. What is widely depicted, and thus repeatedly experienced by people in moving through the diverse sections of a city or, for that matter, the diverse sections of a nation, easily comes to assume a kind of universality. Favored images, when repeated over wide areas and in diverse settings, reverberate powerfully, signifying apparent truths if for no other reason than the frequency of their repetition. The beliefs

FIGURE 36

Billboard in suburban Cincinnati, Ohio, 2000. The ad seems to say that home ownership in a quality neighborhood is highly prized by "Belongers" and "Emulators," but especially by "Achievers."

they deliver are made to seem not only self-apparent, but everywhere accepted and acceptable. Conversely, when a social group is rarely depicted in signs anywhere, it is substantially marginalized in public consciousness.

SIGNS AND TERRITORIALITY

The posting of one's own signs is a means by which individuals or groups of individuals can advertise themselves and enhance their sense of belonging. Americans, like people in other societies, post signs in two principal ways: through personal appearances and through the marking of territory. The first involves gestures and other body movements (body language) as a kind of person-focused communication. The second involves taking jurisdiction over space well beyond the body. Whereas advertising signs speak impersonally to a general audience, personal territorial markers speak more directly to others. Whereas advertising signs imply social control, territorial markers exercise social control directly. At work is human territoriality.

Territoriality involves any mechanism that promotes two things: personalization of an area or space and the defense thereof. A place is first rendered distinctive by its would-be owner (or owners) and then defended through one or another behavior.[34] Thus, territoriality involves a range of strategies calculated to establish degrees of access, both person to person, and person to thing.[35] This implies a form of geographical classification (body space, personal space, interactional territory, public space, etc.), but territoriality is also a form of communication that emphasizes human separation, jurisdiction, and responsibility set in spatial context. Sign display — the actual marking of the body or of a geographical space with signs — is a commonly employed strategy. Signs are posted not just to mark territorial boundaries, but to discourage territorial encroachment through violation (unwarranted use), invasion (the physical presence of others), and contamination (the rendering of a space impure with respect to its use).[36] Signs, through written words or pictorial symbols, make explicit what many other territorial markers only imply.

Signs worn on one's person certainly serve to amplify individual persona.[37] They are part of the individual's self-presentation. They serve as signifiers of the signified self, part of the "front region" erected for purposes of social encounter.[38] Tattoos, badges and patches, clothing inscribed with signlike messages, and actual signs held in the hand or strapped to the body: all are commonly encountered in public places. A police officer's uniform carries numerous indicators that he or she is sanctioned in taking jurisdiction over public space and in maintaining order in private places as well. But much of the "signed" clothing seen in the United States today carries only personal messages, usually of lifestyle implication. Worn especially in leisure time, labeled and often highly decorated T-shirts, sweatshirts, and jackets serve to attract attention, helping the wearer in some way to stand out. By changing such clothing, one is able to switch sign messages, in the process skewing one's persona: what one appears to be to the gaze of others. Of course, "signed" clothing and now, increasingly, body tattoos have become so common in the United States that they have lost much of their attention-getting power.

Signs may also be attached to things close at hand to become part of a person's performance. Especially helpful are things regularly carried, such as a student's backpack. Automobiles greatly extend the body's mobility and thus the body's geographical reach. Being regularly used, cars are frequently personalized through signage, for example, bumper stickers (fig. 37), license plates, and tire covers (fig. 38).

Personal messages are also inserted into public space through the defacing of property. Graffiti usually sends highly personalized messages. Sometimes it is an individual's signature that merely says, "Here I am" or "I was here." Graffiti artists take pride in spreading their work over wide areas, positioning it in visually prominent locations. Inscribing their marks on railroad freight cars, for example, sends their distinctive IDs over broad areas. Graffiti art reached a peak in New York City in the 1980s, some graffiti artists coming great distances to "make their mark" there, especially by defacing mass transit cars. An anti-graffiti task force with a $25 million budget was organized in New York in 1995. In 2000 alone, the effort produced over sixteen hundred arrests.[39]

Graffiti art also serves to mark social territory, most notoriously the home "turf" of street gangs (fig. 39). Gang members reinforce personal identity and status among their peers by marking areas of intended gang influence.[40] Offensive graffiti often involves alienated individuals or groups crudely attempting to project themselves into public awareness. On the other hand, many illegal signs, graffiti-like in their execution, are posted by individuals seeking only information or help (fig. 40). Graffiti and graffiti-like signs are forms of spatial colonization: individuals or groups colonizing public space for private purposes.[41]

Home is the most personalized of all places. It is where the individual feels most fully self-assertive. It is where he or she maximizes pri-

FIGURE 37
Bumper stickers in downtown Philadelphia, 2000.

FIGURE 38
A "Cornhuskers" car in Omaha, Nebraska, 1994.

vacy through the fullest assertion of territorial control. At home, entry by others, one's ongoing activities, and furnishings in support of those activities directly reflect personal prerogatives. Owning a home, as opposed to renting or leasing one, is highly prized in America as a form of social success. The feeling of being "at home" may be extended to a neighborhood or to other places that a person frequents.[42] The landscaping of one's yard may serve symbolically to connect people with neighbors or to separate them. So also does the posting of signs, especially those at doorways or driveway entrances (fig. 41). When quest for privacy fully dominates, however, what usually gets asserted is not neighborliness, but property rights: the "no trespassing" sign ubiquitous to America.

Traditional valuing of the nuclear family in America has brought an extraordinarily large number of signs to the American scene. Prime among them are markers rarely thought of as being signlike. Some residences in America are named, identified as distinctive places, following the British custom of house labeling. But nearly every house in the United States does carry a number that, in following postal regulations, orients it directly to a public street or road. There are probably more street address signs in the United States than signs of any other kind. Cemetery markers represent another kind of sign little thought of as such. Cemeteries provide finality, with graves made into "resting places" idealized as territory eternally possessed. Although the fads and fashions of grave marking have changed over time, the impulse remains the same: to celebrate lives past. That celebrating increasingly

FIGURE 39
Gang graffiti in a Chicago neighborhood, 1973.

FIGURE 40
Sign seeking lost property in Providence, Rhode Island, 1998.

involves highly personalized messages, if not from those interred, then from those surviving (fig. 42).

Individuals and other private entities (such as corporations) act territorially through the posting of signs. But so also does government, public signs serving not only to regulate people's behavior, but to assert, in general, powers of public territorial jurisdiction. Public signage not only

FIGURE 41
Farm mailbox near Millersville, Illinois, 2000.

informs, but also seeks to persuade and exhort. Traffic signs, for example, demand compliance in directing action. As previously discussed, motorists are informed of speed limitations, turn prohibitions, and parking restrictions, for example. Sometimes posted signs blur the distinction between public and private. In the United States, many powers

FIGURE 42
Grave marker near Teutopolis, Illinois, 1990.

of regulation are legally extended to private individuals, enabling them, for example, to regulate activities on private property. Figure 43 shows a sign located along a back alley in downtown Indianapolis, its assertion of property rights enhanced (rather than blunted) by a wry sense of humor communicated in an official-looking sign format.

■ Signs, we have emphasized, are an important means by which Americans have come to conceptualize themselves and others. Advertising signs help to outline the dimensions of our modern consumer society, every sign potentially a mirror in which to reflect who we are and who we might become. Beyond the reading of signs, the making and posting of signs also influences personal identity, especially through assertions of human territoriality. From marked clothing to signed automobiles to posted real estate, Americans have learned to express their individuality. Most of this activity, however, follows normative patterns of expectation, such signs being easily anticipated and acted upon.

The use of signs is so commonplace as to be easily taken for granted. Nonetheless, every sign one confronts implies a fundamental line of questioning. Whether an advertising sign or a private or public territorial marker, the kinds of sign emphasized in this chapter, sign users are always faced with the same important decisions. Should they or should

FIGURE 43
Sign posted along an alley in downtown Indianapolis, 2000.

they not comply? Should they or should they not conform? Whatever the pattern of response, responding (as when responding to any challenge) becomes a test of personal identity. Does the sign apply to me? If so, to what extent? Accordingly, what is the proper reaction? Signs of all kinds challenge. All signs seek response. From such challenging, and the patterns of behavior thus fostered, comes much of what we consider ourselves to be.

PART FOUR **SIGN AESTHETICS**

7 : Signs and Landscape Visualization

Signs attract attention according to layout, labeling, color, and all other internal characteristics. They are designed and positioned specifically to be noticed: to capture the gaze of passersby in ways fully impelling. Consequently, signs tend to detract from other things potentially seeable in landscape. As signs serve to orient, inform, and persuade, they are intended to cue either immediate behavior or to predispose eventual behavior. And they do this always with place implication, giving meaning to and taking meaning from geographical context. Given their impelling nature, signs can variously reassert, modify, or altogether change the meaning of things seen nearby. Thus, signs contribute forcefully to the "reading" of landscape as a kind of visual text. What signs contribute to landscape visualization and to the interpretation of built environment is the emphasis of this chapter.

As visual display, landscape serves to cue meaning. Everything seen stands potentially to be interpreted, and interpreted, more importantly, as "sites" around which one can organize social interaction with others. Places are recognized centers of expected satisfaction or dissatisfaction: situations or settings where, based on some kind of prior experience, something positive or negative is anticipated. Places, as centers of meaning, are seen to nest in landscape, sometimes in a hierarchical manner. They take meaning according to behavioral intention: what a person is doing and what a person wants to do, coloring his or her "take" on situations at hand. It is through place-orientation that landscapes are known.

Perhaps it is by actually moving through built environment, place to place, that landscapes are best known. Movement requires what some scholars call "way finding" (sometimes called navigating), where place-orientation, especially in regard to location, is absolutely essential. Some signs serve directly to orient pedestrians and motorists to meaningful places, cuing (even directing) turning behavior, for example. Some signs instruct as to how places are to be used. They assign degrees of jurisdiction to different people or to different categories of people. But where other environmental cues, such as architectural features, also serve to communicate place-appropriateness, signs, through the strength of written words or articulated graphics, tend to do so forcefully and with less room for misinterpretation. Thus, buildings, even fine pieces of architecture, are often covered with signs.

It is quite appropriate that a book concerned with sign history consider, if only in outline, the public take on what signs contribute to landscape visualization. We enter, therefore, the domain of visual aesthetics to ask: What, in the American experience, has been most widely asserted about signs and their visual impact on place? Specifically, what has tended to draw the most criticism? It is quite clear that beauty does, indeed, reside in the eye of the beholder. But Americans, as a society, have adhered quite consistently over the years to a relatively few value positions, both pro and con, in assessing the visual impact of signs as an aesthetic element of built environment. Sign critics have persisted in their felt need to strictly regulate signage, many kinds of commercial sign being singled out as aesthetically demeaning in landscape. Others have vigorously resisted such thinking. Perhaps more telling, most Americans have remained quite indifferent.

LANDSCAPE VISUALIZATION

As a topic of study, landscape invites at least three highly related, but nonetheless different, orientations. There is the landscape of spatial practice: the real-world surround in which individuals move in going from place to place. There is the conceptual landscape: the spatial environment imagined in terms of past place experience. Potentially tying the two together is the depicted landscape: real landscapes and places as represented symbolically in words or in pictures. These three ways of thinking about landscape mirror one another. People come to real places with preconceived notions: ideas, in part, obtained from words and pictures. These ideas, in turn, predispose comprehension by encouraging the seeing of some things over others. They stand, it might be said, as windows of awareness through which visualization is directed. Attention, of course, is a precondition for fuller comprehension; comprehension, in turn, is the basis for one's actually being in real places.[1]

Substantially, the mind anchors attention in verbal thought. In language, the structural descriptions that represent knowledge might be called "propositions." A proposition is a predicate with a truth value, and a predicate, in turn, is a representation that specifies the relation among a set of elements.[2] The function of attention is to create propositions. Those operative in any one instance are both numerous and varied: propositions in identification, propositions in search, propositions in noticing, propositions in thinking. In assessing a scene, multiple takes always come to the fore in competing lines of interpretation, some propositions focal and others merely referential as background.[3]

Propositional "takes" play out in cognitive systems: patterns of thought relatively stable in memory that, accordingly, stand available to be repeatedly relied upon. These schema (or schemata) enable individuals to maintain degrees of consistency in moving through life from one situation to another. Schema are not rigid, but are open to revision as new situations are encountered, especially new situations that invalidate expectations.[4] Psychologist J. M. Mandler proposed five effects of place schema in the ordinary processing of information.

> First, objects are coded more easily when presented in an organized, familiar scene. Second, well-organized scenes are recalled better and for a longer time. Third, when recalled, disorganized scenes are modified into more usual scenes. Fourth, anomalous objects and violations of physical laws are properly recognized. Fifth, schema-relevant information is more accurately remembered than schema-irrelevant information, which is in turn better remembered than schema-opposed information.[5]

Things in the visual environment, therefore, stand variously as schema-expected, schema-compatible, schema-irrelevant, and schema-opposed. When one encounters a place, one activates the place schemas considered appropriate. One then verifies the presence of schema-expected elements (those elements derived from already activated schemas) before deciding whether a place is or is not an instance of some particular place category.[6] Attention is then devoted to the schema-expected elements of a scene, which can then be more deeply internalized. According to geographer John Pipkin, expected schema influence which objects in landscape are seen and remembered, thus providing a framework for new information, integrating episodic information, guiding the retrieval of information, and determining what information is communicated with others.[7] Although verbal language informs knowing, not all consciousness is verbally discursive. Some landscape propositions, according to Pipkin, are brought to and held in consciousness in the form of quasipictorial surface images that stand, in his words, as "image-knowledge."[8]

Inspired by economist Kenneth Boulding's book *The Image*, planner Kevin Lynch studied what it was that people saw in cities, an effort that culminated in his now classic book *The Image of the City*.[9] Boulding was concerned with what a person believed to be true, his or her subjective knowledge or "image" of the world: that which was visualized by the "mind's eye."[10] Lynch, for his part, tied such concern directly to landscape as visual environment. What was it that made landscapes "imageable"? What stood out to attract attention and be remembered? Lynch classified the visual elements of landscape as follows: paths (trajectories along which people moved and from which they saw and conceptualized landscape), edges (lateral elements seen as bounding potential movement), districts (areas seen as homogeneous in some way), nodes (places, defined at various scales, through which people saw themselves as potentially moving), and landmarks (objects noticed in passing that stood out as points of visual reference).[11] Another planner, Donald Appleyard, built on Lynch's ideas in focusing on urban architecture. He asked why it was that some buildings got noticed as landmarks and others did not. Legibility, he hypothesized, was a function of a building's visual attributes: its silhouette (figure/ground contrasts), size, shape, surface veneer (including styling), and, important to our discussion here, its signs. He also thought legibility was a function of a building's context as measured by: "viewpoint intensity" (the number of people regularly passing by), "viewpoint significance" (its presence at important decision points or points of transition in a city's circulation system), and "immediacy" (its distance and centrality in normal viewing).[12]

Of special interest were the mental thought patterns, or schema, that enabled people to identify and relate to places directly. Behavioral geographer Reginald Golledge distinguished between three kinds of knowledge in that regard.[13] Declarative or landmark knowledge is knowing what is in an environment: objects, persons, and activities variously cued visually. Procedural or route knowledge is associated with problem solving, especially that of "keeping the path" in movement. It involves developing a travel plan and heuristics to translate such planning into spatial activity. Configuratorial or survey knowledge involves notions of angularity, direction, and distance from which integration in an overall frame of geographical reference can be inferred. The latter adds associational and relational components to declarative and procedural knowledge, enabling formation of what geographers have come to call "cognitive maps." At work are propositions clustered so as to inform geographical awareness. Cognitive maps, as schemata, might be thought of as analogous to the real maps a pedestrian or a motorist might use in moving from place to place.[14] They are held in the head, however, rather than in the hand.

WAY FINDING

Signs affect landscape visualization primarily as they attract attention along paths of movement. Before exploring this idea, however, allow us to more fully consider just what it is that people orient to in moving through landscape. In conceptualizing how it is that people visualize landscape, most scholars have thought in terms fully analogous to snapshot photography. Seeing is treated as an infinite set of successive instants, each instant producing a static image in a sequence of images.[15] But James Gibson has argued that the eye, unlike a camera, has no shutter. The eye scans, sweeping over what it sees to produce a sustained image with progressive gain and loss at its leading and trailing edges. No succession of discrete images occurs in one's looking around.[16] Thus could architectural critic Ian Nairn see place as a "continuous thing, a sequence of related parts." Indeed, looking at landscape (at least at visually stimulating landscape) is, he said, like watching a strip tease. What one reads in identifying places is "a matter of alternate tease and strip," with a constant "tension between yes and no."[17]

Way finding is a dynamic, step-by-step process by which people search the visual environment for cues helpful in keeping on course toward some intended destination. Therefore, it involves validating expectations using mental schema organized as a cognitive map. In its broadest sense, way finding involves propositions for coping with route choice, directional change en route, estimates of distances traveled, and, of course, recognition of one's travel goal. As the eye scans landscape, one identifies objects and focuses vision, if only momentarily, on things potentially relevant. Regardless of the nature of serial vision, most scholars agree that visual images so obtained are held in short-term memory until translated into memories of longer duration.[18] Through focus of attention, things seen take on place meaning as they signal some potential for behavior and, accordingly, some implied gain or loss in satisfaction. In motoring through a city, for example, it may be nothing more than recognizing when and where to turn.

Obviously, signs substantially facilitate way finding. Traffic signs not only label routes by name, but serve to point motorists in appropriate directions, establish speed limits, and announce distances yet to travel. As we have previously established, traffic signs are highly standardized and are placed systematically along routes of travel thus to variously alert drivers to situations ahead, define those situations in terms of real world absolutes, demand appropriate behaviors, and even, on occasion, offer reinforcement for decisions correctly made. They are direct aids to anticipating, making, and validating travel behavior. But other kinds of signs can also serve to orient movement. Signs, in general, can serve to visually reinforce architectural and other environmental cues useful in way finding. Buildings passed are made more noticeable, and the place implications of their form, size, or styling made more explicit. But signs can also be distracting, drawing the motorist's attention, for example, away from driving and travel decision making. Figure 44 shows how the main highway through a small Kentucky town turns abruptly at the courthouse square, as route signs at the right indicate. But the motorist's eye is carried ahead; movement is invited down the hill and away toward a distant vanishing point on another road. Advertising signs, especially on the left, also tease. They mark businesses as places of intervening opportunity: invitations to stop rather than encouragement to keep on moving.

Geographer Jay Appleton argued that landscape is visualized through a search for prospect and refuge.[19] Our biological inheritance as a species, he asserted, reflects primitive humans as both hunters and hunted. As regards landscape visualization, we have two innate needs: the need to survey out and over great distances (so as to see prey and

FIGURE 44
Highway at Booneville, Kentucky, 1987.

better anticipate our own danger) and the need to identify places of safety (should danger, indeed, be encountered). Appleton's is a "hazard" approach to cognitive mapping. Translated into modern terms, people seek the opportunity to look expansively: to enjoy uninterrupted vistas of spatial depth if not breadth, the term *vista* implying one's looking forward. And looking forward, in turn, implies the potential to move ahead, vision being stabilized around some point of forward focus. Vistas are variously bounded. The vista along a city street, for example, is confined by the bordering facades of buildings. Figure 45 diagrams what a typical motorist might see in moving down a city street at 30 miles per hour. In open countryside, a motorist can enjoy a full 20-degree field of clear vision, and a 140-degree field of peripheral vision. But along city streets, fields of view are very much confined, as illustrated. In the city, it is along street margins — in the peripheries of vista — that refuge is usually found. Motorists can pull over to the curb to stop outside the traffic flow, or they can turn into parking lots. In busy downtowns, sidewalks offer pedestrians safety from vehicular traffic, and, in turn, the entrances to storefronts offer the opportunity for individuals to escape the crush of pedestrians. Storefront signs not only attract attention in defining stores as specific places of business, but they also announce how those places can be entered: how private space and public space interconnect.

In motoring, the car itself confines viewing, the windshield becoming a kind of lens through which surrounding landscape is seen and comprehended (fig. 46). So also does vehicular speed frame what lies ahead. With higher speed, a driver's concentration on the road and its traffic necessarily increases. His or her point of concentration (or point of focus) recedes, peripheral vision shrinks, and, as attention to nearby detail diminishes, the ability to estimate distance is impaired.[20]

A few studies have sought to explore just what it is that motorists actually see of landscape while motoring. Planners Donald Appleyard, Kevin Lynch, and John Meyer studied freeway drivers in Boston, using cameras to capture eye movement and questionnaires to test recall. Things seen tended to be straight ahead or forward at the oblique. Attention was strongly focused on the road itself, especially on moving traffic at points of decision. But when traffic permitted, sight was also directed, for short periods at least, to scanning landscape at a distance. Landmarks tended to be seen in clusters rather than singly. Very important was the sense of motion. Where objects seen were far away, few, featureless, or moving with the vehicle, the sensation produced was one of floating, or of little forward momentum. Conversely, when attention

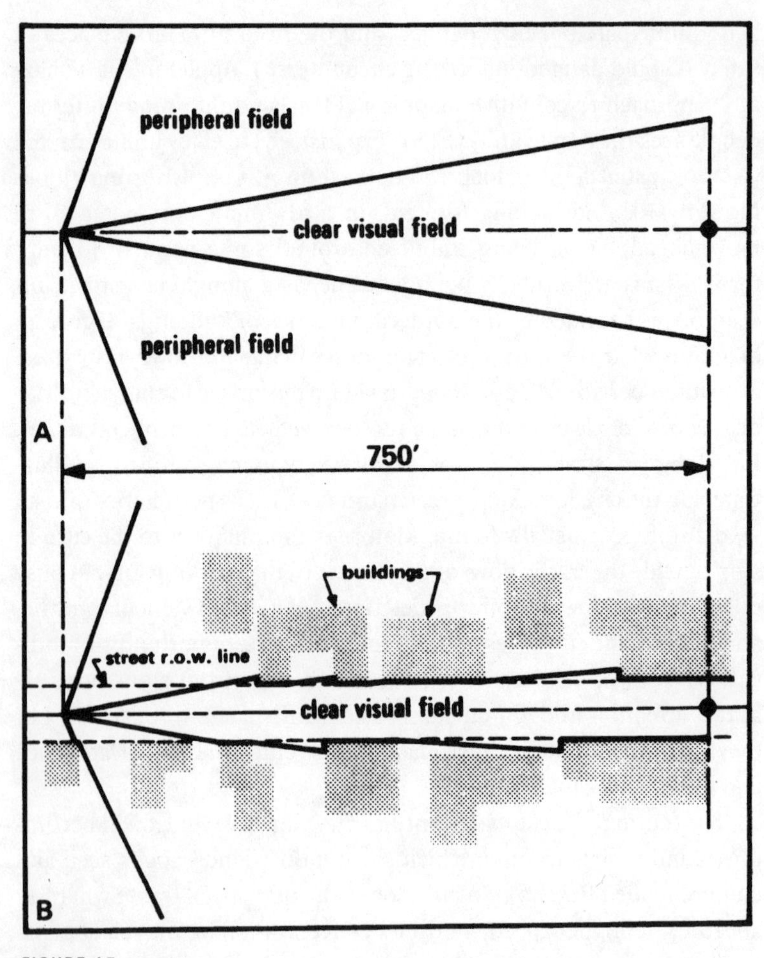

FIGURE 45

Theoretical and actual visual fields at 30 miles per hour. Source: Modified from Leslie S. Pollock, "Relating Urban Design to the Motorist: An Empirical Viewpoint," in William J. Mitchell, ed., Environmental Design Research and Practice, *p. 11-1-4.*

was directed to things nearby, the sensation produced was one of great velocity. Road alignment was important in that it, more than anything else, predicted movement, the geometry of the road standing as the compelling attention-getter. Objects welled up and fell behind, seemingly breaking up as they passed overhead, slipped sideways, or otherwise rotated out of sight.[21] Noticed and remembered were selected billboards and other advertising signs, especially at turns in the road where they intercepted driver lines of sight.

What motorists saw from rural roads was found to parallel closely city findings: the road ahead, road traffic, and traffic signs accounted

AT SLOW SPEED — 30 MILES PER HOUR — THE DRIVER'S EYE IS FOCUSED A FEW HUNDRED FEET IN ADVANCE OF THE CAR, AND HIS RANGE OF VISION INCLUDES THE EDGE OF THE PAVEMENT NEAR-BY. THE ROADWAY SEEMS WIDE

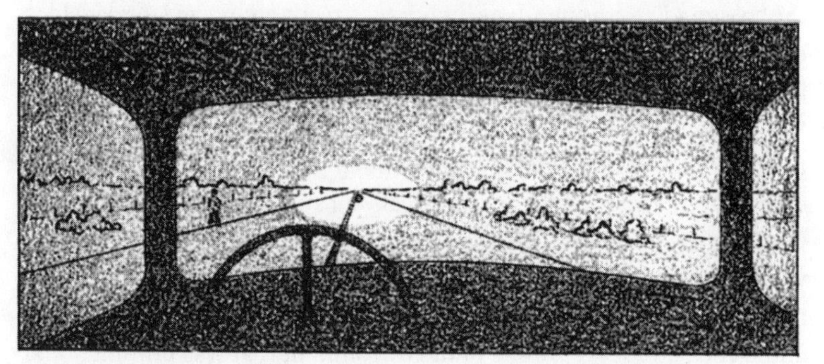

AT FAST SPEED — 70 MILES PER HOUR — THE DRIVER'S EYE IS FOCUSED FAR AHEAD OF THE CAR, AND HIS RANGE OF VISION IS VERY LIMITED. THE ROADWAY SEEMS NARROW

FIGURE 46
Impact of speed on field of vision. Source: Jac L. Gubbels, American Highways and Roadsides, *facing p. 23.*

for a large proportion of a motorist's eye fixations.[22] Translated into highway design, "internal harmony," rather than "external harmony," in the words of Christopher Tunnard and Boris Pushkarev, loomed most important. They defined internal harmony as the way a road composed visually as integrated continuity of form. External harmony was the way a road related visually to its surroundings. Thus, the interstate freeway, they asserted, derived a certain "beauty" from four elements: first, from the harmonious rhythm of its curves (their form, their scale, and their coordination in three dimensions); second, from the proportions of the shapes seemingly enclosed as seen from the driver's seat (especially those of the median strip, formed in perspective, and those between the horizon and the pavement); third, from the way pavements fit into the total "sculpture" of landscape (the way they seemed to cling to hills, jump over valleys, or wind along bodies of water); fourth, from the vistas offered (from broad hilltop views to views confined in valleys).[23] It was the highway engineer's task to design road alignments as continuous, free-floating three-dimensional lines properly proportional and consistent in scale.[24]

By the 1970s, however, new freeway construction in the United States had produced roads that were, perhaps, too free-floating. What Todd Snow called the "new road" was too much a matter of internal harmony. First, wide rights-of-way seemingly isolated motorists from surrounding landscape. Rather than things crowding in close at hand on roads, they were kept at a distance, even beyond peripheral vision. With the "old road"—the two-lane highways directly connected to roadsides of commercial and other activity—the motorist could not ignore his or her surroundings. With the "old road," Snow wrote, motorists were "in" the landscape and fully interactive with it, but with the "new road," the motorist was embedded in a confined pattern of behavior so specific as to be "insulated from practically all surroundings."[25] The freeway, as mere extension of car and driver, represented, in the extreme, what geographer Edward Relph called "simple landscape": built environment that declared itself openly, presented no problem or surprise, and lacked subtlety.[26] It was built environment that was unifunctional in its simplicity.

Perhaps freeways are overly predictable, driving behavior being too easily anticipated. Freeways were built, after all, for the sole purpose of moving people in cars at high speeds. Therefore, they sustain a visual environment that is totally rational but also largely passionless. More than ever, motoring has been made a means of merely getting somewhere; travel by car has become a means rather than an end in itself. Not only is the passing scene much diminished, but higher speeds produce a kind of "straightening" or "stretching" of path, the path thus being made all-important. "Haste, which causes a fleeting assessment of environmental factors," wrote ethologist Fred Fischer, "leads to a sensorially determined 'straightening.'"[27] Even before freeways became common, American motorists seemed to suffer from an obsession with "making time."[28] It was, in journalist James Flagg's estimation, the inescapable impulse to keep going: a "perpetual automation" that "befitted a nation known for rocking chairs and chewing gum." The national bird, Flagg had concluded, even in the 1920s, was not the eagle, but "the squirrel in the cage."[29]

THE ROLE OF TRAFFIC SIGNS

Succinctly stated, traffic signs may be classified as regulatory, warning, or guide signs. In order to be effective, they must fulfill a need, command attention, convey a clear simple message, command respect, and be placed to allow adequate response time.[30] In normal driving, especially on freeways, the motorist confronts an enormous flux of visual information. Distractions can be many. Traffic signs exist to allow motorists to quickly refocus attention on driving and on navigation. Traffic signs must be designed and placed so as to be conspicuous not only individually, but as they relate together in giving driving instructions. First, signs must stand out against their backgrounds. They must be noticeable. Second, sign messages must be unambiguous, and they must be concise for rapid reading. Signs, according to their purpose, must be positioned in ways readily anticipated, with the same kinds of messages displayed in like situations so as not to confuse.[31] For drivers moving at different speeds, permissible viewing time varies. Motorists moving at thirty miles per hour (forty-four feet per second) have about eight seconds to see and to read traffic signs, but at sixty miles per hour they have only four seconds (fig. 47).[32] Traffic signs must be carefully programmed, standardized as to appearance and systematically positioned.

Sign legibility involves such variables as style of lettering, size of lettering, and contrast of lettering on a signboard's background.[33] Uppercase letters are easily recognized, but tend to be read individually rather than as constituted words. Use of both upper- and lowercase lettering, on the other hand, tends to facilitate the reader's interpreting of phrases. Rule of thumb dictates that the width of a letter's vertical stroke is best kept at one-fifth its height. Also, words should be central in a signboard, and moved to the right or to the left only when combined with a pictorial symbol. Black letters on either yellow or white backgrounds are, perhaps, the most legible, followed by such combinations as yellow on black and white on blue. However, background color, besides providing contrast for lettering, can also be used to code messages. Thus, along interstate freeways, signs of green background (and white lettering) cue turning and other route-orientation directives. Universally in America, red means "stop" and yellow "caution." So also are meanings coded through sign shape. Thus, octagonal signs symbolize "stop," and diamond-shaped signs "caution." Every American driver is expected to understand such symbolization. When driving, we see and attend to such signs largely subconsciously.

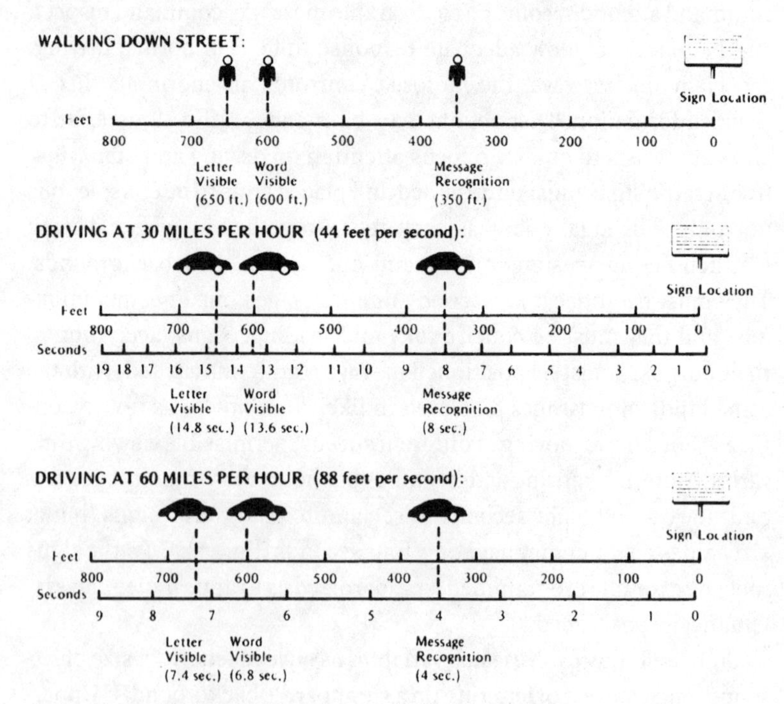

FIGURE 47

Sign readability at varying rates of speed. Source: Karen E. Claus and R. James Claus, The Sign User's Guide: A Marketing Aid, *p. 95.*

In recent years, "icon" or pictographic traffic signs have come increasingly to the fore in the United States, based on European precedents (fig. 48). Since they do not require the motorist to interpret written text, they cue behavior more rapidly.[34] There are three kinds of glyphs or pictorial symbols in use: object-related, concept-related, and abstract. Object-related glyphs look like what they represent: for example, the outline of a man working or the profile of a truck descending a hill. Concept-related glyphs suggest how something works: arrows depicting the separation of opposing traffic, for example. Abstract glyphs, as the name suggests, have less immediate relationship to their referent: two lines crimped together to symbolize the narrowing of a

FIGURE 48

Text and icon signs contrasted. Source: Theresa J. B. Kline, Laura M. Ghalli, and Donald W. Kline, "Visibility Distance of Highway Signs among Young, Middle-Aged, and Older Observers: Icons Are Better Than Text," Human Factors *32 (Oct. 1990), p. 613.*

road, for example. Such signs epitomize their genre as traffic regulators. While conveying essential information, they remain largely unobtrusive until needed.

As traffic signs facilitate movement, they inadvertently reinforce the road: the path as a dominant in landscape visualization. They function primarily at the edge of driver awareness along the sides of streets and highways. True, they are noticed as needed. But they are also noticed when they do not work especially well, when signs do not fulfill our expectations and thus confuse the meaning of place accordingly. Traffic signs instill degrees of consternation when they are worded ambiguously, provide too little or too much information, or stand in conflict with one another or with other kinds of signs.[35] Lack of signage perplexes, leaving motorists uncertain as to what to do or where to go. Traffic signs can also annoy for their redundancy or frequency of encounter. Wrote William Saroyan in appraising motoring in 1960s America: "There are too many numbers and signs along the highway. We know too well where we're going."[36]

THE ROLE OF COMMERCIAL SIGNS

If traffic signs are intended to reinforce movement on streets and highways, advertising and other commercial signs are not. Rather, they are intended to take advantage of movement. Their purpose is to divert the pedestrians' and the motorists' notice away from moving. As landmarks, they may aid way finding, but that is not why they are erected. It is not the purpose that they are intended to serve. Commercial signs are not as readily classified as to function. We can only say that they variously identify, orient, inform, and persuade. Perhaps they are better thought of in terms of structural type: electric sign spectaculars, billboards, and storefront signs, for example. Commercial signs affect the seeing of landscape in at least two fundamental ways. First, they assertively impinge on visual awareness to either enhance or confuse landscape as visual display. They can add variety to make apparently empty scenes more interesting visually. In great densities, however, they can also make landscapes seem overly complex and even chaotic, sometimes with positive effect, but often without. Second, they stamp the seeing of landscape with commercial implications. They enhance the consumer's sense of place in fostering shopping and buying habits. As planner J. M. Bennett observed of outdoor advertising: "Few people really realize the mental effect of things which are apparently passed unnoticed. Billboard advertising . . . [for example] stamped on the subconscious mind by repetition."[37]

The English writer, J. B. Priestley, in traveling in the American Southwest in the 1930s, was invigorated by advertising's landscape implications. "Here along these western highways, with their fine surfaces, careful grading and banking, elaborate signs, their filling stations, the auto camps, their roadside eating-houses and hotels, their little towns passionately claiming your custom in a startling sudden glare of electric signs, is a brand-new busy world," he wrote. "Gas," "Eats," "Cold Drinks," and other messages were spelled out in paint by day and in neon light after dark. "There was rapidly coming into existence a new way of living — fast, crude, vivid — perhaps a new civilization, perhaps another barbaric age — and here were the signs of it, trivial in themselves, but pointing to the most profound changes, to huge bloodless revolutions." There was no real planning, he said, only business people trying "to make a few dollars." But one could envision the change eventually sweeping over the world. "This new life was simply breaking through the old like a crocus through the wintry crust of earth," he mused. "And it was here in America . . . that the signs of it were most multiplied and clearest."[38]

Traditionally, in big city downtowns and out along commercial thoroughfares, commercial signs were positioned to intercept the eyes of oncoming pedestrians and motorists. They served as focal points in anchoring vistas along streets and in enhancing sense of place, whereby business locations were differentiated one from another. Corners were prized for billboards and electric sign spectaculars, especially intersections where traffic stopped, thus lengthening viewing time (fig. 49). Such signs were fully exclamatory in their intent, featuring short messages and vivid graphics, often animated. And they could be, especially in large numbers, very entertaining. Nowhere were advertising signs more stimulating, as we have said, than in New York City, especially along Broadway at Times Square. On visiting the city in the 1930s, two Russian humorists, Ilya Ilf and Eugene Petrov, wrote: "The electric parade never stops. The lights of the advertisements flare up, whirl around, go out, and then again light up: letters, large and small, white, red and green, endlessly run away somewhere, [and] only a second later . . . renew their frantic race."[39]

Advertising signs enlivened traditional commercial streets, especially as signs of different eras intermixed to suggest a historical depth of time. During the Depression of the 1930s, Farm Security Administration

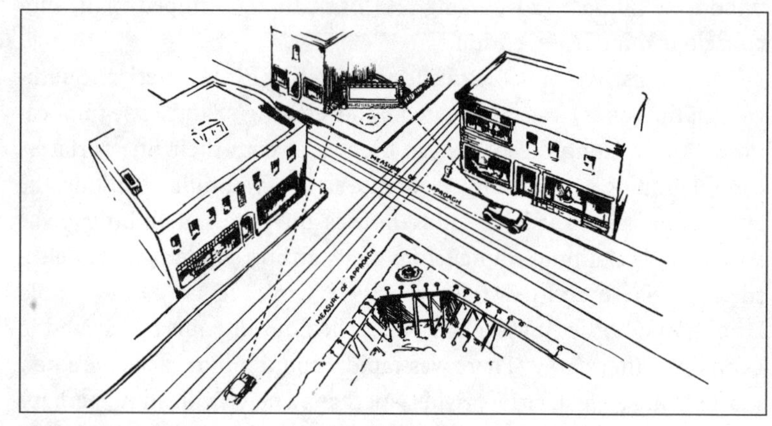

FIGURE 49
Ideal location for a downtown billboard. Source: Wilmot Lippincott, Outdoor Advertising, *p. 273.*

(FSA) photographers, charged with depicting America, especially its rural and small town landscapes, regularly focused on signs as a means of representing the nation's essence. Advertising signs were made to speak for human connections that went well beyond commercialism. Many of the photographs, being widely reproduced, were absorbed into the nation's collective memory, their images standing today as icons of the American past.[40] FSA photographers seemed to say that places, through their signs, assumed not only identity, but personality. Signs implied human energy. They humanized landscape as communication. Landscapes that lacked signs, especially commercial districts in cities, appeared, as they still appear today, underenergized, substantially empty of activity, and less interesting accordingly (fig. 50).

In many ways, advertising is parasitic upon place. Advertising, for its assertiveness, often obscures other social discourses anchored in landscape.[41] Indeed, ads typically occur together with, or are embedded in, social narratives to which they make no direct reference or with which they, indeed, conflict. An ad sandwiched into the middle of an essay in a literary magazine, for example, is somewhat akin to a billboard sandwiched between buildings on a downtown street or positioned strategically at the curve of a rural highway (fig. 51). Highway planner John Robinson decried the intrusiveness of highway billboards. As roads became straighter and smoother, allowing cars to move faster, billboards "became bigger, more vulgar, more blatantly colored to attract the eye." Billboards, he said, had evolved as "a garish kind of roadside architecture." If not by placement, then by design, they were overly aggressive, and thus, as communications, they tended to dominate landscape and place.[42]

As art triumphed over copy in the 1930s, eye-stopping, if not garish, illustration came to the fore in billboard advertising. Poster layout was kept simple, and, along with a logo prominently displayed, things pictured were made bold, clear, and colorful (fig. 52). Words were held to a minimum. Words in ads only invited difference of opinion, some sign makers claimed, especially when they smacked of hyperbole. And advertisers were, indeed, prone to exaggeration. Pictures, on the other hand, prompted less controversy. They were more readily understood at face value, people tending to "believe what they saw." It was also argued that pictures could surpass written copy, both in their ability to intensify emotion and in their capacity to communicate several emotions at once. Visual imagery could also reach illiterate consumers, conveying "dense messages" to them despite their lack of education. Evocative pictures upped the consumption ante by inducing the customer to obtain fullest satisfaction by "buying the ad" along with the product.[43]

Signs, commercial signs included, speak as landscape signature, inscribing built environment with meaning. Through signs, landscapes

FIGURE 50

Downtown street in Durham, North Carolina, 1991. Commercial streets largely stripped of traditional signage stand starkly symbolic of downtown decline, visual vacancy replacing visual vibrancy.

FIGURE 51

Billboard placement, good and bad. Source: Wilmot Lippincott, Outdoor Advertising, *p. 274. As a rule of thumb, an eye-level billboard should not be angled more than 20 degrees from the normal position of facing oncoming viewers. When the angle must be greater than 20 degrees, the width of the lettering should be increased.*

FIGURE 52

Idealized poster layout. Source: Patrick Knox Smith, "Laws of Layout in Display Art," Signs of the Times *87 (Oct. 1937), p. 48.*

are personalized and humanized by being made more intensively a means of communication. For most Americans, however, landscape, as built environment, has traditionally been thought of not so much as a process of communication as a mere physical surround, a container variously useful in the conduct of life. Landscape stands as structure variously functional, with some kinds of landscape considered more functional than others. And, for some Americans at least, some landscapes are seen to be more attractive than others, especially as they encode social values other than those strictly commercial. The valuing of visual aesthetics in landscape has generated highly contentious debate over the years. And signs, especially commercial signs, have, as often as not, been the focus of such debate.

SIGNS AND LANDSCAPE AESTHETICS

Commercial signs, as they proliferated in the American scene, excited negative reaction from categories of offended citizenry, especially cultural elites who preferred to see landscapes of refined beauty rather than of elaborated commercialism. Signs, especially in high densities, were viewed as offensive. They stood as a nuisance. Advertising signs especially were seen as visual blight, a contagion that virulently spread to infect places.[44] What underlay sign blight, of course, was the exaggerated embrace of materialism: America's quest for goods and services made available in a substantially unregulated capitalistic marketplace. Central to this situation was the valuing of land as a basic economic resource and the reluctance to interfere with private property rights. With the moving frontier (where American society, many commentators felt, was given form), nature was exploited in the creation of agrarian rurality and industrial and commercial urbanity, utilitarian values, rather than aesthetic values, were forcefully asserted. But as the nation matured, degrees of tastefulness, sustained by elite cultural proclivities, brought order and style to architecture and landscaping, at least selectively. Veneers of gentry civility were laid down in the nation's cities and small towns. Valuing the picturesque was clearly manifest in big city parks, civic centers, and upscale residential districts, especially in the new suburbs that sprouted in the late nineteenth and early twentieth centuries. Emerging, or so it seemed to some, was a nation of landscape refinement.

Blight, as geographer David Lowenthal reminded, implies retrogression from a previous state of grace or, as we might add, from an expected or hoped-for future state of grace. For a landscape to be blighted, he argued, blight needed to be fundamental. It needed to affect the entire built environment and not just a piece of it. It also needed to be long-term in its effect and not merely ephemeral.[45] Critics charged, for example, that commercial signs indiscriminately located in great numbers along commercial streets were inherently distasteful. Such signage countered the ideal of the "city beautiful": the city of parks, boulevards, and elegant, if not monumental, architecture. It also countered pastoral ideals, rooted in agrarianism, brought to city suburbs as a means of countering or correcting urban ills. But with widespread improvement of highways nationwide, critics turned their attention more and more to the automobile roadside. The highway billboard was made to epitomize sign blight. Signs threatened to blight the countryside and its pastoral purity. If, in an aesthetic sense, cities represented utopia unfulfilled, rural America would soon become utopia lost.

Critics of big city downtowns and of small town Main Streets saw most commercial signage as intrusive. Most Main Streets, argued Harry Lucht (of the Architects League of Northern New Jersey), reflected little more than a "sign maker jamboree." They created a carnival atmosphere. Signs tended to be shoddy, and, in their shoddiness, they diverted attention from storefront design and from other serious aspects of environmental quality. "Many property owners spend huge sums for architectural fronts," he wrote, "and then permit the occupants to destroy their attractiveness with ugly signs."[46] Activists seeking to "beautify" cities, through sign regulation and other environmental controls, worked through local garden clubs and other, mainly women's, civic groups. At the national level, they found voice through such organizations as the American Park and Outdoor Art Association and its successor, the American Civic Association.

Billboards, as a blighting pestilence, they said, had left the nation's cities and marched out into the countryside, arraying themselves for the motorist's view. Scenery stood diminished by, if not obliterated by, the sign intruders. After 1920, new organizations were formed and charged with preventing the continuing visual assault. The National Council for Protection of Roadside Beauty, for example, published a magazine, the *Roadside Bulletin*, which appeared irregularly through the early 1930s. The *Bulletin*'s articles, most of them lavishly illustrated, outlined the threat of an unrestrained outdoor advertising industry. Verse was commissioned for the antibillboard crusade.

> Our Country, tis of thee,
> Great land of billboardy;
> Of thee I sing.
> Land of the tourist pride,
> Land of the hot dog fried;
> On ev'ry mountain side,
> Big signboards swing.[47]

Distributed to be displayed by members, both on cars and, interestingly enough, along roadsides, were signs that read: "Billboards Offend Tourists Who Spend," "S.O.S.: Signboards Obstruct Scenery," and "Billboards Hide the Country's Pride."[48] How paradoxical. Signs, it was asserted, should be used to combat signs.

Numerous were the newspapers and magazines that, although not themselves advocating sign regulation, allowed sign critics to have their say, especially through the publishing of cartoons (fig. 53). In the 1930s, many authors conducted inventories, driving highways to count offending signs. They tabulated not only billboards, but "snipe" and other signs as well. In Georgia, in the fifty-five miles between Gainesville and Decatur on U.S. 23, there were ninety-five signs "plastered" on barns. In the fourteen miles of U.S. 41 between Atlanta and Marietta, there were 234 signboards, 64 of the them of the "largest type." These signs appeared, in other words, sixteen to the mile or one every six seconds when motorists drove at the fastest allowable speed.[49] But numbers, in and of themselves, did not win arguments. Nor, for that matter, did appeal to raised aesthetic sensitivity. Appeals to the American public to boycott products advertised on billboards received little support. And few Americans listened even when signs were criticized as hindering safe driving.

Of course, advertising signs posted close to highways were a hazard. They could not help but divert driver attention from the road ahead (fig. 54). The *Roadside Bulletin* asserted: "There has come a growing realization that the large and lurid billboard, placed to attract the attention of the passing motorist, is likewise distracting his attention from the important business of driving at high rates of speed now permitted on trunk highways in the open country."[50] In 1957 the Pennsylvania Roadside Council published a circular entitled *This or That? (Scenery or Signery?)*. Motorists were admonished to investigate for themselves the billboard danger. "Outdoor advertising is designed to attract attention," the pamphlet read.

> Thus it is obvious that it must distract attention to accomplish its purpose. Make your own safety test! How quickly can you pick out an official traffic sign or route marker from a forest of billboards and other signs? How many feet do you travel with your eyes off the road while reading a billboard? Remember, at 60 miles per hour it takes 292 feet in which to stop your car.[51]

Of course, the outdoor advertising industry stood vehemently opposed to sign regulation. Industry trade associations took the lead in

FIGURE 53
Antibillboard cartoon. Source: Reprinted from the Chicago Evening Post, *in the* Poster *14 (Dec. 1923), p. 18.*

countering antibillboard and other sign-control initiatives. Lobbyists worked state legislatures and Congress. Company spokespersons covered city councils and county boards whenever restrictive ordinances were proposed. Industry publicity offices distributed their own brochures and saw to it that articles opposing sign regulation circulated in popular magazines. The din of sign advocacy, of course, reached its

FIGURE 54
Antibillboard cartoon. Source: Roadside Bulletin *2, no. 4 (c. 1933), p. 5.*

highest pitch in sign industry trade journals. Sign advocacy proceeded along two trajectories: discrediting opponents as being outside the American mainstream and crediting the sign industry with an essential centrality.

Prosign advocates sought to characterize the advocates of sign regulation as wrong-headed, disparaging them, for example, as fanatic crusaders. "I wish them one and all a long life and a dreary one," wrote the editor of *Signs of the Times*. "The billboard flutter is just one of the minor species of hysteria which are overspreading the land." Signs, he said, were not in any way "destructive." At worst, some signs might be "obstructive." Some signs might obscure a scenic view here or there. But that was little price to pay for outdoor advertising's pervasive good. And even when signs were ugly, their ugliness could not compare with real environmental degradation.

> How can one view with complacency our commercial country, our tunneled, quarried, undermined and deforested hills, our wretched architecture, not to mention overhead wires, telegraph poles, our railroads, mills and factories, and other far more permanently unpleasant things, and then reasonably and consistently single out for mad, petty antagonism the busy little sign board that never did any harm in the world but advertise?[52]

Prosign advocates emphasized outdoor advertising's positive contributions to American society. Was not the billboard the essential handmaiden of commerce? To think otherwise, it was argued, could even be construed as un-American. "Standardized Outdoor Advertising is so thoroughly a part of the fabric of American industry that it cannot possibly be removed without danger to the entire structure," concluded one journalist writing in the *Poster* in 1928. It was only "the Bolshevik type of mind" that condemned the constructive role that outdoor advertising played in marketing goods and services.[53] As Congressman James C. Wright argued some forty years later: "Jobs depend upon the creation of new business and the expansion of old business. But these factors depend directly upon our ability to market and sell the growing volume of products which this places on the American market." "Outdoor advertising," he concluded, "is therefore a vital and necessary part of . . . economic expansion without which this nation would be in serious trouble."[54]

Industry apologists claimed that roadside signs were definitely not a safety hazard. Quite to the contrary, they argued, signs added interest to landscape and were thus entertaining to motorists who might otherwise find highways (and highway driving) monotonous and boring (fig. 55).[55] Not only that, but billboards could actually add beauty to landscape. Most posters, being designed by leading graphic artists, were visually pleasing, if not beautiful. Arrayed along highways, they stood gallery-like, to interest and to excite the "millions of people who never even think of visiting an art museum."[56] Billboards on the roadside, in other words, were "the people's art gallery." Poster art was a legitimate part of American culture, it was argued. And, as an art form, it was worth preserving as part of the nation's heritage. Rather than obscuring scenic beauty, as critics charged, billboards, as often as not, hid unsightly things, especially in cities. Signs brought diversity to landscape, rather than the dull sameness that sign regulators sought to impose. Overall, signs "energized" landscape by bringing "vibrancy to the built environment not easily provided by other means."[57] It was

FIGURE 55
Prosign billboard. Source: "Iowa OAA Goes All Out to Win Friends, Serve Public," Signs of the Times *150 (Dec. 1958), p. 41.*

pointed out that those who thought signs to be ugly, and thus demeaning of landscape, often used devious means in arguing their case. "One of their favorite tricks," wrote the Missouri Sign Association's Richard Newman, "is to find a very bad part of town with old buildings, crooked telephone poles, and signs that are ancient and in disrepair. Then they use trick photography to make five or more blocks of signs to appear one directly on top of the other."[58]

Sign regulation was made out to be a form of censorship. As John Ise argued, freedom to communicate was basic to American society. Freedom of speech — the right to be heard — implied, as well, the right to be seen. "The right to communicate visually in the outdoor area — in good taste and within the law — would seem to be one of our essential freedoms," Ise wrote.[59] Rather than impose censorship, it was better to allow the sign industry to regulate itself. Did not the codes of the Outdoor Advertising Association of America (OAAA) provide sufficient protection against sign excesses? That organization precluded posting signs (1) so as to create a hazard to traffic, (2) on rocks, posts, trees, fences, or barricades, (3) on streets facing public parks where surrounding streets were residential, and (4) at locations that obscured important natural scenic beauty. Members were prohibited from tacking, pasting, tying, or erecting cards, panels, or signs of any description anywhere except on structures that conformed with association guidelines.[60] What else would reasonable people want?

■ Sign industry apologists wonder at the intensity of sign criticism. But the critics, sometimes described in the 1920s as "long-haired professors" and "short-haired women," cannot be dismissed as mere fanatics or as inconsequential "do-gooders." What signs contribute to the visual environment is, indeed, an important concern. It strikes at the heart of what the nation's geography ought to be and what that geography ought to look like. Did not the sign industry itself claim that signs, commercial signs especially, are essential to the nation's economic functioning? Signs are important not only for the profits they generate for sign makers (the true bottom line in the sign industry's quest for laissez-faire), but are especially important as things seen and attended to. That is how they affect the economy: they stimulate consumption through persuasion. Nothing else in landscape, except, perhaps, other people, so readily and so consistently captures the average person's gaze. And signs do so because they stand as surrogates for people: people communicating messages in absentia, usually anonymously, for anonymous others who might see and be persuaded. Landscapes are known from paths used in moving. Signs, for their part, are intended to divert pedestrians and motorists from seeing and attending other aspects of landscape. Signs "foreground" themselves in turning other things into background. Signs tend to be noted first, and other things in landscape are noted thereafter. No wonder the advocates of the picturesque and of the beautiful, in opposing the mundane of the commercial, focus criticism so vigorously on signs.

What sign control enthusiasts envision are landscapes signed less for commercial purpose and more for social purpose. If highways, for example, were largely devoid of advertising signs, motorists would be left to traffic control signs and, accordingly, would be better positioned to appreciate natural and other scenery. Landscape vistas would loom up ahead and occlude behind, uncluttered by messages difficult to ignore and largely irrelevant to the more important purposes of scenic appreciation. Such thinking, of course, ignores the fact that advertising signs themselves might be seen as scenic. Billboard advocates are not entirely mistaken in their claim that poster displays can be attractive as well as attracting. Average Americans are not sensitive to many of the aesthetic tastes cultivated by the nation's cultural elites, and poster displays can resonate and have resonated as "art." Few Americans stop to think about what they see in landscape. Few Americans even use the term. Sign regulation advocates, therefore, have worked to substantially educate the public not only to see beauty in landscape, but to see landscape plain and simple. The public must be challenged to see the visual environment as anything other than mere physical surroundings.

The visual environment does offer cues to place meaning. And it is in and through meaningful places that life is lived. Signs are, perhaps, the most important of all landscape cues, given their assertiveness. The question remains: Just what role should different kinds of signs be allowed to play in defining the places landscape contains?

8 : Sign Regulation

Outdoor advertising has always been disproportionately influential. The physical size of billboards, including the recently devised "jumbos," let alone "snipes," and the duration of any particular sign has been far less than most other architecture, both the buildings and the spaces, along city streets and automobile highways. After one hundred years of refining strategies, the outdoor industry came to share but 1.5 percent of the advertising budgets throughout the nation by 1988.[1] Yet no other element of the city curbside and rural roadside has stirred public exchange like outdoor advertising. Proponents and detractors have often argued their case in spirited terms and sometimes invective. To fully appreciate the stakes that signs manifest in the landscape, some attention must be directed to the debate, now more than a century old.

Our discussion of sign regulation in this chapter will proceed largely on geographical lines. We will first treat sign regulation in the city, where the calls for control first arose, and turn in succession to two other locations. Those will be the roadside and, last, as we name it here, the horizon. *Horizon* here has both literal and metaphoric meanings. In the literal sense, *horizon* connotes the attention large billboards command to look above and far toward the vanishing point between surface and sky (fig. 56). In the metaphoric sense, *horizon* is also used about questions of the future for ordinary signs in all locations, on the road and in the city, which stir heated debate almost everywhere about the future of visual culture. What will the landscape look like if commerce is allowed to have its way in billposting? What will happen if control depresses commerce? Answers from both those for and against sign regulation attempt, with varying successes, to mobilize the political construct of "the people." Both appeals reach far into a receding future. Our approach to sign regulation in this chapter is not wholly a function of spatial relationship, for, in stepping from city to roadside to horizon, we proceed chronologically as well. Those are the places on which successive debates on sign regulation centered. Yet place will set the tempo, because no eschatological progression toward either the goal of laissez faire capitalism or communal hegemony can be demonstrated to be at work. People interacting in place and time to evolve society in all its complexity often reveal steps back from as well as toward the alternative objectives of those contesting for and against sign regulation. What

FIGURE 56
Antibillboard lobbyists charged that Missouri had three times more billboards per mile than eight surrounding states and in 2000 barely lost a vote to ban billboards in the state. This view was made on I-70 near Temple City, Missouri, in 2001.

people want their business, neighborhood, or highway to look like — that remains the constant in the exchanges about sign control.

THE CITY

Merchants posted signs on America's urban landscape for several centuries until, in the late 1800s, conditions engendered efforts at regulation. Industry thrived. Population exploded, and America's cities rapidly extended over larger areas, with the consequence that severe demands descended on services there. Coupled with industrialism's new-found wealth, these conditions enabled some to realistically ask how to elevate the quality of life. After a century and a half, dating from the earliest American cities, it is hard because of the mixed results today to imagine reformers' initial enthusiasm for solving the problems of urban life. Aesthetic concerns shown brightly in the agenda of the City Beautiful movement.[2] Contents deemed crude and corrupting, for example, those of theater posters luring youths to portrayals of the white slave trade, emboldened the forces against billboards, but contents were not the lone problems.[3] Signs, but especially billboards, placed exclusively at billposters' and their customers' discretion, offended the sensibilities of the self-appointed arbiters of urban taste. Often themselves the beneficiaries of considerable wealth, those workers for the City Beautiful ideal judged unbridled commercialism to be out of step with the strong pastoral undercurrent they shared with American culture.[4] Vibrant colors and animated designs, including electric lights, contrasted with the urban reformers' subdued visual values. Parks, or reserves importing the look of the natural countryside into the city, were essential to the proposed therapies for redeeming congested and diseased urban living.[5] Signs placed only with calculation for where they might most effectively advertise goods and services contradicted the essence of reserve. Planning how a city should evolve, rather than letting it grow in response to innumerable decisions for individual advantage, was central to City Beautiful thinking. Opportunistic billposting for private gain was anathema to the vision of orderly progress. Denouncing O. J. Gude's electric spectaculars for the H. J. Heinz Company in the public epicenter of the nation's most famous city, one defender of Manhattan's visual honor wrote, "In the evening the dancing flash-lights of the '57 varieties' of beans, pickles, etc. thrown in the face of all who throng Madison Square — the real center of life and art of New York — are unimaginable except in nightmare."[6] Their offense to good taste stood but a short distance from the Dewy Arch, the type of Beaux Arts monument that the reformers preferred. Shouldn't the public appreciate the contrast, even if Heinz and Gude did not? The same writer lamented public indifference and lectured at length about the need to teach and uplift public taste.[7] J. Horace MacFarland, the movement's best known lay voice, refrained in public from denouncing billboards altogether, because they did have economic merit, but privately wished them abolished.[8] Contemporaries denounced the City Beautiful movement for its elitist presumption of superior taste.[9] Here is a dilemma that has run throughout the demand for sign regulation in America's past and not only in the City Beautiful phase: how to persuade a democratic people that their populist values are inadequate and should be enlightened?

William Wilson has traced the public pundits and urban designers leading the City Beautiful movement's attack on billboards.[10] The legal results can be followed in pivotal court cases building principles through various precedents about sign regulation since the late nineteenth century. Those results amounted to a substantial bulwark for sign regulation but without fully satisfying the reformers.

In the 1890s and for a decade thereafter, several cities adopted but had ordinances overruled that regulated the height, size, position, and location of billboards, including distance from the street. Two cases illustrate the fortunes of early municipal sign regulation. An ordinance in Topeka, Kansas, stipulating a setback from the street five feet more than the height of the sign and an ordinance in New York City limiting "sky signs" to nine feet were overturned in their state supreme courts, respectively in 1893 and 1909.[11] The opinion of the New York court typified the reasoning. "Does the ordinance, so far as it relates to sky signs, come within the police power, or is its purpose simply to prevent or restrict a lawful business which, it is alleged, has been extended until it has become offensive to good taste?"[12] Failing any demonstration that the signs threatened public health and safety, the court struck down the ordinance.

The most exceptional break in the courts' rejection of aesthetics as sufficient by themselves for upholding sign control legislation in the early twentieth century occurred in the Philippines, an American protectorate since the Spanish American War. The Philippine high court in *Churchill and Tait v. Rafferty* in 1915 produced some of the strongest statements in any American court, before or since, for sign control. The plaintiffs charged that the Philippine Attorney General's power to remove billboards according to a statute that the Congress ratified in 1915 was unconstitutional. Both parties agreed that the billboards presented no indecent information and that their structure did not threaten public health and safety. The defense for Rafferty, a Collector of Internal Revenue in the Philippines, however, stated that billboards were "offensive to the sight,"[13] and on that basis alone the billboards could be removed in accord with the statute of 1915. Churchill and Tate countered that Rafferty had committed a "taking," that they were deprived of property without the due process of law guaranteed under the Fourteenth Amendment of the United States Constitution.

The court decided that the "police power" inherent in every government to regulate individual activities for the common good extended to the aesthetics of the specific billboards in question.[14] Although the billboards did not discomfort or inconvenience people, they affected people's primary aesthetic sense, namely, sight. Although advertising was legitimate because it was "a cause and effect of the great industrial age through which the world is now passing," "beautiful landscapes are marred or may not be seen at all by the traveler because of the gaudy array of posters announcing a particular brand of breakfast food, or underwear, or the coming of a circus, an incomparable soap, nostrums or medicines for curing all the ills to which the flesh is heir, etc., etc." Billboards should not prevent people in "search of outdoor pleasure" from enjoying natural landscapes. Further, billboards do not gain their advertising value from their position on private property but rather "upon the use of the channels of travel used by the general public." Regulating billboards is consistent with the acknowledged right of governments to regulate "streets and other public thoroughfares."[15] Acknowledging that they were out of step with many of the highest state courts at the time, the Philippine justices nonetheless unanimously agreed that any advertisements deemed "unsightly" fall within the capacity for regulation.[16]

Nothing so broadly stated and consistent with the City Beautiful reformers would be confirmed in an American court again until mid-century, long after the City Beautiful movement had ceased. The decision bolstered none of the concurrent cases for sign regulation on city streets. Might the United States Supreme Court's dismissal of an appeal on the case, however, have reflected sympathy for the aesthetic argument in the nation's highest court, as one analyst suggested?[17] A legal scholar at the time of the case, fully aware of the currents shifting in favor of various urban reforms, saw in *Churchill and Tait v. Rafferty* sound reason to believe that "in the course of time American courts will reverse their decisions and frankly include aesthetics among the subjects for which the police power may be properly exercised."[18]

The ardor for sign regulation did not dampen. A journalist in Chicago exemplified the strong sentiment for cleaning up cities having grown too quickly without heed for their appearance and held outdoor advertisers especially guilty. "Surrounding every vacant lot or building in the process of erection are billboards in the most awful combinations of colors."[19] Parks and boulevards, some of the City Beautiful's crown jewels in urban reclamation, were special victims of unregulated signs in the eyes of the disgruntled, often because those signs were displayed within view of the neighborhoods where the reformers lived.

Cities across the nation enacted ordinances to control signs fashioned and placed exclusively with commercial considerations. Their record in a succession of court cases publicized in legal circles proves handy for searching out the pockets of reform at the same time as it unanimously affirms the courts' general unwillingness to extend the police power to aesthetics. In some of the most celebrated cases, ordinances were overturned between 1893 and 1910 in Topeka; Atlantic City; Chicago; Boston; Passaic, New Jersey; East San Jose, California; and Denver.[20] A passage in the decision against the Passaic statute of 1903 — requiring that no billboards be higher than eight feet above the ground and not less than ten feet from the street line — earned repeated quotation in subsequent decisions against regulation: "Aesthetic considerations are a matter of luxury and indulgence rather than of necessity, and it is necessity alone which justifies the exercise of the police power to take private property without compensation."[21]

Many, including the courts, despite their resistance to sign regulations, acknowledged the potency of visual culture in the new urban boom. Still, the visual realm remained unsubstantial contrasted with property rights and lingered in this strangely anomalous status until rendered in tangible terms. The first step occurred in St. Louis when the city debated elaborate plans for remodeling the city along lines of the City Beautiful agenda. The Civic League, behind the reform, persuaded the St. Louis City Council to enact a sign ordinance in 1905 and in 1907 published a pamphlet whose contents argued against questionable visual culture in visual terms, illustrations of what were believed to be the most offensive billboards. In 1911 the St. Louis plant of the huge Gunning outdoor advertising company challenged the city's sign ordinance in the Missouri Supreme Court. The ordinance limited height (fourteen feet), construction (at least four feet between the sign's lower edge and the ground), setback (at least fifteen feet from the street line, six feet from any building or lot line, and two feet from another billboard), and size (not more than five hundred square feet). Gunning enjoyed a substantial trade in St. Louis, the nation's fourth largest city, more than six hundred signboards displaying 360,000 square feet and leased from six months to three years, depending upon the client. The ordinance levied a fee to permit the construction or repair of complying signboards and triggered Gunning's case when its requests for permits were refused. The Missouri Supreme Court reasoned elaborately from a long list of relevant cases throughout the nation and, in passing, characterized outdoor advertising in an uncomplimentary manner that resonated beyond legal circles. Billboards and sky signs in general were characterized as cheap and intentionally ephemeral. Unprecedented factual evidence was brought before the justices. St. Louis officials testified that behind the billboards thieves hid, prostitutes worked, people relieved themselves, and weeds and debris were allowed to accumulate. Aesthetics were again denied as sufficient reason to legislate against outdoor advertising, but the police power was finally exerted in this well-publicized case to restrain advertisers and sustain the St. Louis ordinance. The *St. Louis Gunning Advertisement Co. v. City of St. Louis et al.* landmark encouraged reformers and jurists alike to explore other dimensions of the police power to regulate signs.[22]

Chicago, home to Thomas Cusack, aggressive founder of one of the nation's most extensive early billposting services, went through a succession of court cases ending in 1917 that further alerted the industry to the growing resentment of traditional business practices. Two frequently cited precedents arising in Chicago in 1905 and 1911 had struck down the city's sign regulations respectively as "mainly sentimental" and a violation of the principle that an individual's rights, in this case to do business, sufficiently protect the public's aesthetic welfare.[23] In their wake, Cusack decided to challenge the city's sign ordinance of 1910 that billboards could not be constructed in blocks facing each other that are half or more residential on both sides of the street unless the owners granted their written consent. Cusack won in a lower court (Cook County), but the city appealed the decision. Relying, understandably, on the concepts that had invalidated past sign regulations — that of a taking and that billboards did not threaten public health and safety — Cusack reasonably expected another favorable decision when the case came to the Illinois Supreme Court in 1914.[24] The Illinois justices in *The Thomas Cusack Company v. The City of Chicago* looked carefully at the city's new evidence that the neighborhoods in question, because they were residential, were not as fully policed or protected from fire as other areas in Chicago, and they concluded that the ordinance was valid. In passing, they distinguished billpostings in residential settings from billpostings elsewhere, on buildings and fences. The United States Supreme Court, to which Cusack appealed after defeat in Illinois, asserted the distinction in upholding the Illinois decision in 1917.[25]

Accepting this distinction meant a discriminating concept of place gained further legal credibility. It had been upheld in numerous instances before with ordinances proscribing certain locations for noisy or foul-smelling businesses. City Beautiful advocates had long tried to assert that for each urban section there is an appropriate and distinctive design and decoration. In elite settings, weed control, fence maintenance, and the "general care" of the vacant lots where obnoxious billboards tended to appear, that is, the total ambience, should be safeguarded. A patrician neighborhood "should guarantee that the billboard will not scream its message across the quiet scene."[26] But the City Beautiful movement had seldom challenged sign posting in poor and working-class neighborhoods and all but expired when Cusack battled the Chicago city council between 1914 and 1917. Courts had upheld regulations by the end of the 1910s but never pronounced aesthetics alone sufficient cause for control. The police power, on which sign regulations were justified for public health and safety reasons, ultimately proved weak footing for identifying place types accepted nationally. Decisions were still rendered on the peculiarities of each case and place.

At the end of the era vindicating sign regulation in the cities, the courts nonetheless, in the words of one analyst, had regrettably relied on "subterfuges." Decisions concealed the right of landscape beauty behind the necessity for public health and safety. Courts were reluctant to analyze the visual power outdoor advertisers wielded to influence the emergent sense of place in the new neighborhoods.[27] In a utilitarian aesthetic tradition, beauty seemed a poor reason for legal decisions. Outdoor advertisers, not only bidding for business but appreciating the long-term advantage of wide public support, knew what the courts were slow to realize and utter in the rapidly evolving consumer culture — signs equal power. Who would formulate the nation in the urban and industrial and, then, the roadside era? What directions and alternatives would signs offer? Vision was indeed a most practical matter.

Outdoor advertisers alloyed self- and civic-interest in their growing public awareness as they grasped for more than client satisfaction. The fledgling *Signs of the Times* early exhorted the industry not to display in the most controversial of landscapes, namely, scenic areas. From the summer of 1909 through the winter of 1910, the Associated Billposters and Distributors of the United States and Canada posted advertisements free of charge for the Anti-Tuberculosis Association's national campaign. During World War I, outdoor advertisers figured prominently in the federal government's crusade for fuel and food conservation and the war bond drive. Government generally acknowledged outdoor advertising's astounding capacity to mobilize public opinion. Women (whose recruitment to the wartime workforce strengthened their independent buying power), bankers (slow to accept advertising's role until convinced by the war bond drive), and returning veterans (who benefited from jobs that advertising stimulated): those groups aroused signmen to a fuller awareness of their own potential, like shamans of a new spiritual force.[28]

Outdoor advertisers approached their rights vis-à-vis the public with increasing sophistication. Many local firms had never clashed with authorities over local ordinances,[29] and some heads of big firms, notably Edward A. Stahlbrodt in Rochester, New York, Barney Link in Brooklyn, and Kerwin H. Fulton in New York City, argued for inoffensive texts and discreet display sites, against the practices of billposters doing whatever they thought profitable (fig. 57).[30] Their voices combined with the outdoor medium's acceptance among the various advertising media plus the legal victories of the City Beautiful proponents to persuade the Outdoor Advertising Association of America at its first meeting to codify self-regulation principles for the entire industry. In 1925 the association not only undertook voluntary censorship of those members displaying "false, misleading, or deceptive" claims but specified that posting sites irritating the public should no longer be used. Off-limits

FIGURE 57

The Thomas Cusack Company helped promote tasteful billposting and maintenance throughout the industry by the 1920s. This illustration from the Cusack Company appeared in a publication on outdoor advertising in 1923; it compares the company's restraint (below) with the previous work of an unidentified billposter (above). Source: Wilmot Lippincott, Outdoor Advertising, *p. 7.*

to members of the association in good standing were either structures or bills that created traffic hazards, were daubed or tacked, invaded streets or parts of streets that were purely residential, or were in other locations where the resentment of reasonable persons was justified, namely, on streets facing public parks in residential neighborhoods without the owner's approval and into "natural scenic beauty spots."[31] The association's official spokesmen saw in those policies certain proof of their industry's maturation. Regulation, self-imposed and ordinance-enforced, had helped redefine cities, but temptations lay ahead as opportunities dawned along the new roadside.

THE ROADSIDE

Rival pressures for and against sign regulation swept quickly onto the roadside in the 1920s, when construction of the nation's all-weather highway system got under way. An outdoor advertising executive in Ogden, Utah, was stimulated to consider remodeling his billboards and maintaining them in a parklike setting after participating in a local college debate on sign control and appeased women's clubs by removing three painted bulletins on a highway entry into the city. Beautification became an industry stratagem at the same time reformers kept pushing their agenda. The dialectic moved the regulation issue to the brink of a declaration for the complete prohibition of roadside signs. Attempting to quiet opponents by elevating practices beyond legal requirements has long risked problems for industry reformers. Some signmen believe it will only encourage proponents of sign control to point out that the industry can always do better regardless of its own improvements.[32]

Roadside blight in the form of opportunistically placed displays and poor art displays became an issue for Elizabeth Boyd Lawton, who attracted a national network of women's clubs from the 1920s through the 1950s. She corresponded with them, conducted sign surveys, and publicized the campaign of her National Council for the Protection of Roadside Beauty by means of the *Roadside Bulletin*, published between 1930 and 1950. Billboards represented for her an assault on the open countryside by the commercialism of urban centers, from which she sought relief in country retreats (fig. 58).[33]

State statutes and court decisions effected change meanwhile. Beginning in 1913 in Hawaii, the Outdoor Circle, an influential women's club, organized a campaign against billboards and, by 1927, bought out the islands' lone outdoor advertiser and removed billboards throughout the islands. Hawaii started before the coming of the automobile. In 1925 Maine adopted a law providing for the removal of signs along the automobile roadside where they obstructed visibility at highway intersections, endangered motorists on the highway, or were constructed without consent of the owner on whose property they stood.[34]

Massachusetts spawned a peculiarly strife-ridden chapter in the history of sign regulation, one that drew national attention. It ended in 1935 with important consequences for city streets as well as rural highways. After 1905, when the state's highest court overturned the state capital's regulations of sign size and location in *Commonwealth v. Boston Advertising Company*, smoldering reformers only grew more determined

FIGURE 58
Lawton's crusade against billboards charged that this example, lighted at night, "is a distinct traffic menace." Source: Roadside Bulletin *2 (May 1933), p. 3.*

and raised the issue before the entire state in the constitutional convention of 1917–1918. As a result, the new state constitution included the amendment that "advertising on public ways, in public places and on private property within public view may be regulated and restricted by law."[35] In 1919 Massachusetts's highest court affirmed that outdoor advertising could not be banished but could be restricted.[36] Municipalities and the state itself set about adopting and enforcing sign ordinances, with the predictable result that grievances arose. The *Roadside Bulletin* focused attention on Massachusetts's growing court cases in the controversy.[37] With suits accumulating since 1925, Massachusetts's highest court combined fifteen total cases in 1931 and four years later rendered the long-awaited decision in *General Outdoor Advertising Company, Inc., et al. v. Department of Public Works*. It proved to be the landmark widely anticipated. The court upheld Boston's refusal to renew a license for a roof sign across from the statehouse, Concord's denial of permits for signs in that historic village, and several actions by the state's department of public works against commercial signs on the highways. While the outdoor advertising industry was competing successfully with other media, including the new radio technology, because signs were unavoidable elements on the landscape, the Massachusetts court turned that unique advantage against the plaintiffs. "It is protection against intrusion by foisting the words and emblems of signs and billboards upon the mass of the public against their desire." Outdoor advertisers "are not exercising a natural right; they are seizing for private benefit an opportunity created for a quite different purpose by the expenditure of public money in the construction of private ways."[38]

Despite the reformers' celebration of the decision in 1935, it nonetheless fell short of the coveted unequivocal indictment of billboards. An analyst in the *Harvard Law Review* would have preferred that the court enunciate the principle that "the ownership of land confers no constitutional right to use it for projecting propaganda into neighboring areas."[39]

The war of words escalated. Both sides sought sweeping victories. Advertising was in its heyday in the 1920s, no claim for its benevolence seeming preposterous.[40] At least one federal highway bureaucrat believed outdoor advertisers were tantamount to claiming a "vested right" along highways.[41] Ever the aggressive voice for outdoor advertising, Foster and Kleiser, which dominated roadside advertising in the West, divided the world into two halves in 1929: the commercial highway and the scenic highway. The latter were "beauty spots" requiring that "every good citizen" neither operate a business nor advertise on them. "In the main, however, highways are used for commercial purposes. . . . As cities expand and business travel increases between them, these businesses — these pioneers of future industry — take their stand along the highways."[42] A roadside imperium was envisioned free for private exploitation and conferring the pursuit of happiness on all who worked and traveled there — an apotheosis of the American dream. California's attorney general in San Francisco in 1928 affirmed that zoning, consistent with the police power, justified urban sign regulation but noted that "such control does not exist along the highways running through the sparsely settled country."[43]

In 1931 the second meeting of the Conference on Roadside Business and Rural Beauty adopted a bill proposing national standards for zoning scenic areas beside the highways. Representatives of the American Nature Association and the National Council for Protection of Roadside Beauty withdrew from the conference rather than support any compromise permitting outdoor advertising along rural rights-of-way. Agreement stalled after the committee deferred a decision until another meeting.[44]

Courts inched closer to an unconditional protection of aesthetics under the police power in several subsequent billboard cases. In *Perlmutter Furniture Company et al. v. Greene, Superintendent of Public Works, et al.* (1932), the New York Supreme Court upheld the state highway authority's power to erect a screen to block a billboard distracting motorists on a sharp highway curve on the approach to the Hudson River bridge outside Poughkeepsie. "If, incidentally, the outlook from the road is improved by shutting off the view of the billboard, so much the better."[45] In *Kelbro, Inc., v. Myrick, Secretary of State, et al.* (1943), the Vermont Supreme Court upheld a statute enabling the state to deny construction of a billboard within 300 feet of a highway intersection and within 240 feet of the center of the road. Quotations from the Massachusetts decision of 1935 and, finally, from the once solitary Philippines case in 1915 were mobilized to deny claims to the outdoor advertisers' right.[46] Yet reformers were unable to claim an unconditional victory for their aesthetic. The justices in the Perlmutter case nicely summarized the status that the aesthetic reason alone for sign regulation had achieved after nearly fifty years of debate over billboards: "Beauty may not be queen, but she is not an outcast beyond the pale of protection or respect. She may at least shelter herself under the wing of safety, morality, and decency."[47] Practical reasons remained preeminent in jurisprudence. Sign regulators and entrepreneurs thus paused in a standoff over the rights of visibility as highway construction came to a halt in the interwar years of the twentieth century.

THE HORIZON

Because the police power often exercised to regulate signs offered no compensation to the advertisers if they were denied the opportunity to advertise, debates in court persisted about sign regulation. Numerous decisions affecting various subjects in the long history of the police power left the limits of government authority uncertain.[48] Given the aesthetic and financial stakes in outdoor advertising, legislating and challenging sign regulation not only continued through the early 1950s in local cases but was bound to spill again onto the national scene with a new program of highway construction.

Opponents joined the issue again in 1955 when Senator Richard L. Neuberger, a Democrat from Oregon, proposed an amendment to the federal aid highway act of that year to grant the secretary of commerce exclusive advertising rights on some rights-of-way for up to five hundred feet from the highway. *Signs of the Times* charged Neuberger with representing an "extremist minority which dislikes outdoor advertising and seeks to eliminate this advertising medium from the American scene."[49] Although defeated initially, Neuberger's provision passed in the 1956 act.[50]

Enactment of federal interstate legislation in 1956 planning forty-one thousand miles of new highway construction stirred forces deeply. Those seeking regulation reasoned that because the federal government was paying 90 percent of the cost of the new construction, it was fair to consider national standards along the right-of-way. Prominent figures such as Robert Moses, a staunch and renowned highway proponent, enlisted in the ranks of regulation and decried the interstate act's omission of sign control. Lobbyist Scott Lucas, former United States senator from Illinois and momentarily a Democratic Party consideration for the vice presidency, argued points in the Senate Subcommittee on Public Roads that were standard among outdoor advertisers: restrict signs and many small roadside businesses would die; restrict signs and threaten the livelihood of more than three million people. After considerable debate among legislators on interstate sign control, Congress enacted a law in 1958 that required the secretary of commerce to define sign standards. What had failed in 1931 seemed about to happen. Four sign types were made permissible, and the states were not required, but encouraged, to participate in adopting them. According to a bonus system, a state that acquired advertising rights along its interstate highways could receive federal reimbursement not to exceed 5 percent of the project's total cost. That amounted to 0.5 percent of each state's allocation on interstate projects where the standards applied.

Opportunities for new profit and the countervailing call for new regulation seemed disproportionate to the following stir. By the secretary of commerce's estimation, the interstate standards would apply to less than 2 percent of the country's entire road and street mileage. The three-quarters of a million miles already built and maintained with federal aid were exempt from the new standards, and about 25 percent of the miles along the new interstate were also exempt. They, however, were not in areas zoned commercial or industrial but were in rights-of-way never before developed.[51]

The highway again loomed as one of the most revered American places — the frontier. What should this place be? Renewed assertions of the old entrepreneurial claim to commercial ownership of the highway pushed the dialogue ever harder. "Highways were not — and are not — built primarily for beauty." "To insert that highways be dainty — at taxpayer expense — resembles the idea of putting green paint on miles of railroad track — to make it pretty."[52] This was *Signs of the Times*'s answer. Polemicists like Peter Blake in his classic *God's Own Junkyard* (1964) reminded readers that Congress had fearlessly debated largely on aesthetic grounds in 1958 and scolded Americans for their pragmatic values that served to "encourage the desecration of this country and discourage those who wish to preserve or (God forbid!) beautify it."[53]

Northeastern, midwestern, and Pacific Coast states comprised the bulk of the twenty-five states that took advantage of the bonus clause of 1958. Southwestern and southeastern states formed the region for which Senator Robert S. Kerr, a Republican from Oklahoma, most effectively argued the traditional outdoor advertising business viewpoint. Of the six states whose compliance was challenged in their highest court, Georgia alone, where land-use controls were not then popular, failed to uphold its compliance with the federal act. Approach of the deadline for compliance with the bonus incentive on June 30, 1965, triggered further debate. The customarily placid *Reader's Digest* incited public rancor with an article entitled "The Great Billboard Scandal of 1960." The article threatened "billboard slums" beside the new interstates if strong measures were not taken, because the outdoor advertising lobbyists, unrepresentative of most advertising's more restrained stance, routinely frightened legislators, farmers, and labor unions into believing that regulation was the goal of a fanatical minority, not the popular will, and would significantly diminish collateral income and cost jobs. In 1961 President John Kennedy amplified demand for regulation with his assertion that billboards detracted from both beauty and safety on the highways and recommended that the bonus be doubled.[54] What would happen when the bonus expired?

Aesthetics gained legal force meanwhile. Some court cases between 1944 and 1964 upheld the exclusion of advertising signs from commercial zones, where they once proliferated, and retroactive agreements for sign removal. The "Vermont doctrine" was suggested. Likely named for the decision in *Kelbro v. Myrick* in Vermont, a milestone in the series of decisions denying an entrepreneurial vested highway right and the strength of aesthetics alone for sign regulation, jurist Ruth I. Wilson explained how the concept of vision as an easement could supplant the often ineffective regulation through the police power.[55]

While many thought of a future in which signs on the highway frontier would determine the look of the landscape, Americans generally began to literally look farther forward and higher overhead as the effects of the bonus began to play out. Enter the "jumbo." No one had foreseen the ineffectiveness of the Bonus Act provision of the 660-foot setback. Because it was believed that billboards were unreadable at that distance, it seemed reassuring to believe that none would be constructed. Jumbos, billboards covering between twelve hundred and twenty-five hundred square feet, occasionally more than five thousand square feet, began cropping onto the landscape beyond the 660-foot setback, however, and they were quite readable from the windows of passing vehicles.[56] The horizon was now dotted with them, like an endless stage set moving ever forward as travelers drove into it.

President Lyndon Johnson, motivated by his wife's strong voice for highway beautification, called in February 1965 for reform of the Bonus Act of 1958 because it appeared ineffective in cleaning up the nation's roadsides. Sign regulation crested at this high water mark in 1965. A White House conference in May 1965 convened eight hundred participants, many of whom rejected any off-premises advertising on primary and interstate highways. Phillip Tocker, chairman of the Outdoor Advertising Association of America, probably seeking to sustain the hard-won recognition of billboard advertising's rightful place in American society and certainly consistent with its policy since 1925 against signs on scenic highways, proposed that signs be permitted only in areas functioning as or zoned industrial or commercial (fig. 59).

Later in May 1965, the president's moderate bill was introduced into the Congress. (The bill was soon colloquially renamed the Lady Bird Johnson bill because of her strong endorsement of it.) Although it upheld Tocker's concession, it also imposed a 100 percent penalty in federal aid highway funds on any state not agreeing by January 1, 1968, to comply and required that by July 1, 1970, nonconforming signs must be removed. Those provisions were to be achieved under each state's police power, and compensation was to be paid advertisers only in states where the police power was not authorized. The billboard lobby split. The American Motor Hotel Association, fearing its members' business would perish or decline without informational signs, and the Roadside Business Association, asserting that the federal government should leave roadside beautification to each state, opposed the president's bill. During Congress's debate, the Outdoor Advertising Association of America was able to amend the legislation to provide mandatory compensation rather than amortization for the removal of nonconforming signs. (Under the principle of amortization, a sign company is given a limited period when earnings can be recouped before a sign not conforming to the law must be removed.)[57]

Opponents of the compromise at the heart of the Highway Beautification Act of 1965 tore it apart in practice. Many state officials and businesspeople never accepted Tocker's conciliations. Although the billboard opponents' strongest proposals had been shunted out of the

FIGURE 59
Phillip Tocker illustrated an article defending the role of good yet appropriate outdoor advertising with this view, claiming its visual chaos: "This Is Not Standardized Outdoor Advertising." Source: Tocker, in John W. Houck, ed., Outdoor Advertising, *p. 47.*

legislative process, they registered again in the widely circulated book *Highway Beautification* by Charles Floyd and Peter Shedd. Its subtitle, *The Environmental Movement's Greatest Failure*, forewarned readers that the 1965 legislation betrayed the high hopes from which it sprang. The list of betrayals revealed how the opponents of sign regulation achieved their ends by inserting loopholes into the legislation. First, the secretary of commerce enforced the 10 percent penalty in only one state for failing to comply with the act; the penalty was simply too harsh for general application. Second, national standards evaporated when an amendment required the secretary to arrange regulations "consistent with customary use" in each state. Third, each state was permitted to determine whether the unzoned industrial or commercial areas could be classified in that category and signs be permitted. Fourth, all states were required to compensate advertisers for signs removed.[58]

Vermont and Maine took their own course in setting some of the strongest controls on outdoor advertising in the wake of the Highway Beautification Act's temporization. Hawaii and Alaska ranked beside them, having enacted a strong code in 1927 and 1959 respectively. All four states found a convincing practical rationale for their remarkable statutes in the need to sustain the pastoral image for tourists, who fortified their economies so richly. Vermont's law of 1968, culminating from a legislative proposal dating back to 1955, has been celebrated as a via media between total sign abolition and total entrepreneurial license. Vermont adopted logo signs for its highways. These directional signs, color coded according to the several roadside services, at once satisfied the small businesspersons' insistence that their enterprises cannot survive without signs and the beautifiers' desire to minimize the number and size of signs. Advertisers paid a fee for their business sign to be posted on a compound sign carrying notice of other businesses at appropriate turnoffs, and no other off-premises signs were legal. In addition, information along the highways was provided in strategically located travel information centers, where promotional literature was concentrated for tourists' convenience. Strong state measures, however, remained a route seldom taken. Rhode Island in 1985 became but the fifth state to ban billboards.[59]

Discontent elsewhere ripened in the 1970s and early 1980s. Rural billboard advertisers especially moaned about the impending loss of business. An independent study in Missouri for the Roadside Business Association of that state, for example, concluded that support for beautification was less than expected and that, if implemented, it would end all highway signs business in Missouri. Beautification stalled when no funds for sign removal were appropriated in 1968 and 1969. In the early 1970s, several amendments conciliatory to the sign industry were defeated in Congress but one in 1974 extending sign regulation to the limit of visibility along the interstates gave two years for compliance. By then, outdoor advertisers had erected so many jumbos that an estimated $77,000,000 would have been required to pay for their removal. In 1978 the outdoor advertising industry lobbied successfully to exclude amortization from the 1965 act, with the result that staggering amounts were owed the industry for sign removal. By 1985 it was widely conceded that the act of twenty years before had not merely failed but had financed an explosion of billboards and of clear-cutting to provide sight of them.[60]

The collapse of national standards for outdoor sign regulation reinvigorated debate on the local level, where partisans have demonstrated anew the serious financial and symbolic issues at root. In 1981 the founding of the Coalition for Scenic Beauty, although it principally vowed reform of the 1965 act, also echoed the long-deceased National Council for the Protection of Roadside Beauty in stimulating grassroots support for sign control.[61] By 1988 the list of governments adopting sign regulations over rapidly growing or redeveloping landscapes was expanding.[62] The police power had proven to be useful as an "unclassified residuum of legislative power" derived from the community "to satisfy a public need in a reasonable way," in the words of one legal scholar, and by midcentury the sign industry's assault on the police power behind aesthetic-based sign regulation had failed.[63] Most courts acknowledged by the 1970s that aesthetics alone were sufficient to check sign posting. The industry then turned to the First Amendment to rekindle their cause in court.

Outdoor advertisers had often cited the right to free speech guaranteed by the First Amendment before 1981.[64] *Metromedia, Inc., v. San Diego et al.* in that year, however, became the first billboard case the United States Supreme Court decided exclusively on First Amendment grounds. It acknowledged that billboards, as did all means of communication, had unique capacities and potential abuses. Between the plurality's claim to have laid the foundation for the "law of billboards" and a dissenting justice's denial, that the result in fact was a "virtual Tower of Babel,"[65] the legal profession cannot be blamed for precipitating rather than quieting debate over sign regulation and laymen for their frequent confusion over the subtleties of the ongoing legal exchanges.

In 1972 San Diego enacted an ordinance to eliminate distracting signs that jeopardized pedestrians and motorists alike and to improve the city's appearance. It resulted in a ban on all off-premises billboards but permitted on-premises commercial, government, and temporary political signs. Those distinctions perhaps seemed fair and lawful based on the legal tradition until 1975 that the First Amendment was not applicable to commercial freedom of speech.[66] San Diego acknowledged that the ordinance would eliminate outdoor advertising in the city, that billboards stimulated sales, and that alternative media were neither adequate or were too costly to serve those seeking outdoor advertising.[67]

No single premise guided the court in striking down the ordinance. Opinions varied. Four justices declared the court's opinion. Two concurred but filed a separate opinion. One concurred in the plurality's opinion but dissented on one aspect of it. Two justices disagreed altogether and filed a separate opinion. As a consequence, a majority of the justices did not sign any of the opinions. They did enunciate six guidelines for sign review, but it was certain only from subsequent cases that the First Amendment inserted new and fractious issues into the effort at systematic discussions about sign regulations. Several legal analysts, indeed, charged that the decision was not helpful, one claiming that it undid the previous clarity emerging from the century-long adjudication under the police power.[68] Dissenting Justice William J. Brennan, Jr., concluded about aesthetics that "little can be gained in the area of constitutional law, and much lost in the process of democratic decision-making, by allowing individual judges in city after city to second-guess such legislative or administrative determinations."[69]

Design review ordinances became the subject of innumerable and heated exchanges in courts throughout the nation. Decisions previously left to local government were welcomed in court. For example, if a sign ordinance was not as fully effective as it could be, it could be held invalid.[70] In *Members of the City Council of Los Angeles et al. v. Taxpayers for Vincent et al.* (1984), the Supreme Court acknowledged a long-standing claim among advocates of outdoor advertising control: "These [aesthetic] interests are both psychological and economic. The character of the environment affects the quality of life and the value of property in both residential and commercial areas."[71] The court's decision upheld a Los Angeles city ordinance denying the posting of signs on public property. A political candidate had hired a sign company to make and hang cardboard campaign signs from utility pole cross wires. The decision reaffirmed the court's long-standing premise that the time, place, and manner of any speech conditions the right of its exercise.[72]

Attention to cases that might earlier have been settled locally combined with the peculiar details arising from local cases to ensure a constant docket of sign regulation cases richly mixing history and geography. In the search for reliable precedents or lines of reasoning to forge a legal tradition on the new ground of free speech vis-à-vis sign regulation, history confronted the essentially diverse geographic sensibilities of people disagreeing on sign regulation in their communities in the effort to determine how those communities should look. It was common knowledge, upon which a majority of the nation's highest court now agreed in the context of First Amendment rights regarding the billboard, that the way a place looks — the visual aspect — is not secondary to but of equal importance with the place's economic life. The dissenting opinion in *Taxpayers for Vincent* concurred with this principle but blunted its practical application, reasserting an older view that "laws defended on aesthetic grounds raise problems for judicial review that are not presented by laws defended on *more objective grounds*" [authors' italics].[73]

Aesthetics remained infirm ground for those judging quantifiability to be the only basis of certainty, therefore the only real realm; the scientific mentality was at fundamental odds with the artistic mentality. Notwithstanding over a century of public education finally legitimating the subject of beauty beyond fashionable salons, many remained unpersuaded. Beauty was whatever anybody said it was, and that left room for wild divergence in a subjective quagmire.

Strategies outlined for fighting local control fuel the controversy endlessly. A recent columnist in *Signs of the Times* heeded R. James Claus's premise that signmen often lose in court because they are unprepared for the specialized law in their particular case and identified three steps to strengthen the likelihood of an outcome favorable to the signmen.[74] Claus, a phenom in the battle of signs, is viewed on the one hand within the sign industry as often personally offensive in style.[75] On the other hand, he is seen as an undoubted champion of the signmen's cause. In Claus we find the polar philosophical opposite of J. Horace MacFarland from early in the century but his kin in personal manner. At first a self-appointed champion, Claus now holds formal status within the sign industry. Ranging over the chronicles of sign ordinance history and staying current with cases across the country, Claus regularly wrote advice on how signmen could most effectively fight for their rights. Compromise was not acceptable.[76] Claus, something of an ideologue, reiterated that not only does commerce have the legal right to advertise but that, without signs, small businesses perish. Only very wealthy business would survive. Claus worried about a rising tide of forces against signs and attributed that to a public trend against "environmental pollution." "Governmental decision-makers who have little understanding of the intricacies of visual communication and the causes of environmental pollution" blundered into ineffective solutions, according to Claus's first monograph in 1971.[77]

By the start of the twenty-first century, both variation and consensus characterized the public dialogue. While increased participation has introduced new viewpoints, it can be inferred from them all that signs are a hallmark of contemporary society. Following the generally heightened awareness about signs due to the national debate over the Highway Beautification Act, planners have proposed model sign ordinances since 1971 in trying to pursue a mutually satisfying compromise among adversaries. The planners' term "street graphics" became standard vocabulary in the formulation of those ordinances. The term was predicated upon the principle that signs are more than objects, that graphics are symbols affecting a landscape system. Model sign ordinances were periodically rewritten to keep them relevant to the latest legal decisions and to benefit both public interests and private enterprise. Accordingly, a good ordinance was thought to result in street graphics that would be compatible with the environment, express the community's identity, but still be legible and appropriate to the businesses they serve. Model sign ordinances have also begun to include a role for "historic" signs. The American Planning Association has stipulated criteria and actively promoted the conduct of local sign inventories as a preliminary step in planning for historic examples.[78]

Sign regulation was a category that cut across many other social causes. Guardians of African American neighborhoods, for example, cited the high incidence of cigarette advertisements there and often proposed their banishment in those neighborhoods. In St. Louis, 59 percent of the city's billboards were in African American neighborhoods, 18 percent were in mixed neighborhoods, and 22 percent were in white neighborhoods. Believing their neighbors victimized, in 1989 St. Louis's African American aldermen spearheaded passage of a law banning new billboard construction in that city.[79] Further defying easy categorization, environmental organizations that might seem ready opponents of sign proliferation, for example, Greenpeace, Sierra Club, and the Izaak Walton League, commonly leased billboards to state their case. A recent article in *Audubon* ambivalently indicted billboards as "litter on a stick," yet conceded that outdoor advertising ranked as one of environmentalism's lesser problems.[80]

■ Limitations provoke disagreements, and talk of limiting signs is no exception. Consumers have learned to take their cues from signs. They direct people to the promised land, be it directional signs to a park or billboards to a better mattress. When signs warn of abortion or drugs, they still stand for betterment, albeit from partisan viewpoints. Regulating signs means limiting in some fashion, and this alone helps explain why signs, those three-dimensional objects, flimsy and ephemeral compared to other aspects of the human landscape, stir controversy so ardently. Appropriate sign content has been important through the years. *What can be said?* The answers became more complicated by the end of the first century of debate over limitations. However, *where* signs were placed had consistently drawn sharp comment for over a century. Where signs were posted helped determine what those places would become. And in a society given to perpetual reinvention in quest of a better life, it is little wonder that what signs promised for becoming should have excited and drawn those with a stake in the future both widely and passionately.

Probillboard forces generally far exceeded the political clout of the proregulation forces. Organized labor, coveting sign work and roadside service work for its members, roadside industries, heavily reliant on roadside advertising, and, above all, the outdoor advertising industry, which contributed considerably to legislators' election campaigns and provided them signs, were mighty foes. Women, on the margins of political power, and other spokespersons for sign regulation were comparatively less influential except when they effectively organized public opinion.

Strategy was not the only determinant. Like the ambiguities of federalism buried deep in the nation's constitutional framework, the answers to sign content and place never achieved final resolution in the nation's cultural life. Two fundamental viewpoints were at odds. Essentially, one school of thought contended that beauty was a knowable, albeit subjective, trait appropriate for public policy. Another school of thought conceded that beauty was knowable only in personal terms and incapable of the wide agreement necessary for public policy. Each school might have seemed at various intervals to have won, but in retrospect, one or the other school achieved only temporary victory. Entrepreneurs stood staunchly entrenched as they posted signs along city streets from the late nineteenth century through the early twentieth century, thanks to the courts' restrictions of the police power. Gradually over the twentieth century, those for sign regulation carried the day until by 1965 they stood at the threshold of controlling the choicest frontier since the West itself — the interstate roadside. How could a society committed to entrepreneurship ignore cries that sign regulation jeopardized the very font of the economic system itself, the small businessperson? Charges that signs on highway margins diminished the quality of life, such as the roadside beautifiers habitually made, seemed questionable to those who saw in that landscape proofs of economic health and everyman's capacity for unlimited growth. Limits suppress. Suppression can kill. Commerce won new First Amendment privilege, but beautification was not overwhelmed.

Debate thrived more abundantly than ever at the local level. Signs told what those who posted them wanted for the future. They tried to persuade others and in the process may have won converts. What they surely confirmed was how essential signs were. Who regulates signs controls the future.

Epilogue

Signs substantially affect social behavior as they sustain, enhance, diminish, or change the meaning of places. Consequently, it is the history of signage as a kind of communication that especially needs telling. Signs, as they implicate in human symbolic interaction, are fundamental instruments of social construction. Signs are an important means of empowerment whereby social agendas are advocated place to place. All human behavior "takes place." And signs, for their part, figure prominently in social life as they underpin place-identification, place-orientation, and place-utilization. Signs affect society's geographical ordering, affecting, as they do, people's spatiality, including territoriality.

Signs seem to be especially prevalent in the American scene. Perhaps it is the nation's economy based in free-market capitalism? Perhaps it is the nation's federal Constitution with its guarantees to free speech? Or, as cultural geographer Wilbur Zelinsky has suggested, it may be the sheer spaciousness of the nation that inclines Americans, as a people, to "bedeck the scene" with all manner of superfluous objects, including superfluous signs.[1] We would argue that proliferation of signs in America is in no small way related to the American people's excessive automobile dependence. As Americans rapidly move, cocooned in cars and trucks, they isolate themselves substantially from one another, making sign messages along streets and highways absolutely necessary to binding American society together.

Certainly, automobility has greatly influenced the size and look of signs and continues to lead directly to accelerated sign innovation in the United States. Today, digital signs are rapidly coming to the fore, not just signs produced on computers and printed out on inkjet and electrostatic printers, but, more importantly, signs displayed electronically with electronic lightbulb and light-emitting diode messaging centers. Even more futuristic are signs that use flat-screen plasma display panels, or liquid-crystal display screens, or even projection cubes.[2]

With such sign technologies, advertising signs, for example, can be constantly changed: messages altered in rapid succession, even across sites sign to sign, all quickly programmed and reprogrammed even from remote locations. This makes it possible to deliver information quite literally in "real time." Gas-plasma screens, which look very much like large, high-definition television monitors, whose brightness equates well with the brightest of neon tubing, are capable of displaying vivid, full-color images: images that project with clarity even over great viewing distances. Expensive to build, they currently make sense only where very large numbers of viewers pass by (where daily effective circulation is very high). As costs come down, however, they stand to literally revolutionize the look of cities, especially after dark.

Already clusters of such signs dominate the strip in Las Vegas, enabling the casinos there to advertise entertainment venues, even to the point of posting hour by hour (if not minute to minute) changed theater and meal prices. More economical, and certainly smaller in scale, is use of Smart Paper, a patented display system developed by a subsidiary of the Xerox Corporation. Smart Paper is a thin, flexible polyester substrate embedded with millions of magnetic microspheres, colored on one side and white on the other, that can spin to new positions as electronic charges are applied. It can display two-color text and graphics with some shades of gray.[3] Look for these signs inside shopping centers and department stores. They may, indeed, drive a revival of display window advertising. Look for them inside buses and mass transit cars. In the future, innovation is sure to emphasize portability: not only signs on wheels (or that otherwise move), but such signs with messages set fully in motion across display panels. Streets and highways may soon be crowded by mobile billboards electronically activated. America's continued dependence on motor vehicles encourages sign portability. Why not insert signs directly into ongoing streams of traffic? Why limit them to the margins of traffic flow?

And what of future scholarship? Change is always a stimulus to historical study: analysis that looks backward in time in search of origins. Change also excites scholarly interest in process. How did various sign technologies evolve? Who was responsible? How were signs intended to work? Toward what objectives? What, in the past, in fact, did work? And, of course, what did not? It is our suggestion that future scholarship on signs combine historical and geographical orientations: that signs be considered fully in changing geographical contexts. Signs tell us more than most scholars in the past have readily admitted. Historians have traditionally mined archival documents in search of understanding. Cultural geographers have preferred to "read" the material culture of landscape for encoded messages, looking primarily at buildings and settlement patterns. Of what comparative importance could posted signs and their short messages possibly reveal? Think again. Is not the posting of and the reading of signs, at base, a form of social negotiation? Are not signs fully implicated in the ongoing social interaction that sums to society? Behind every sign lies social agenda. Does not every sign advocate one or another social outcome? And, in so doing, is not attention diverted from other outcomes? Sign messages represent a kind of shorthand, shorthand for underlying social issues.

In the United States, signs speak especially of choice. Americans like to keep their options open. That is, they want to be free to choose. Consumer decisions are thrust ceaselessly in our faces daily until we sleep, our social acceptance depending, even as we sleep, upon the choices we make. But signs can also exhort basic values well beyond the mere invitation to consume. Whatever enjoinder commands our attention, we confront an invitation to comply. Personal freedom, so basic to America's civic creed, cannot be practiced unless two or more answers are possible for every question. Signs are an important means of posing questions.

In the United States, signs also speak of convenience. Americans like to think that options are immediately at hand. They like to minimize the amount of expense and effort expended in solving problems, however trivial. Signs do more than just lay out possibilities. Signs, indeed, facilitate. They tell what things can do. They tell where things are. They suggest solutions. They offer answers. Americans value success. And what Americans know from signs, especially advertising signs, helps them to measure personal accomplishment, perhaps the most important kind of success. Americans value progress, many valuing mere change for change's sake. America is a land always becoming. Signs are not only a medium for advocating and reporting change, but, as signs themselves change, they bring a certain kind of temporal vibrancy to the fore.

Americans like to feel that they belong. And here is where signs are especially important as they contribute to people's reading of landscape for place meaning. Signs can be used to assert community and do so at various scales from that of the nation down to that of locality. Governments impose their jurisdiction and regulate public spaces through the posting of signs. On the other hand, Americans also like to champion their individuality. And nothing promotes the appearance of personal distinctiveness quite like the wearing of signs on one's person. Signs are used to assert the privacy of home and rights inherent in the ownership of private property. As a form of communication, signs suggest what is and what is not normative. They help set expectations. They validate. They invalidate. They predispose. In no way can full understanding of American society be achieved without comprehending how it is that signs in America work, and especially outdoors in public spaces.

Notes

INTRODUCTION

1. See Herbert Blumer, "Sociological Implications of the Thought of George Herbert Mead," *American Journal of Sociology* 71 (1966), pp. 535–44. For an updated account of the symbolic interaction concept, see Norman K. Denzin, *Symbolic Interactionism and Cultural Studies: The Politics of Interpretation*.
2. E. Gordon Ericksen, *The Territorial Experience: Human Ecology As Symbolic Interaction*, p. 59.
3. See G. A. Kelly, *The Psychology of Personal Constructs*, vol. 1, *A Theory of Personality*.
4. Blumer, "Sociological Implication," p. 535.
5. Denzin, *Symbolic Interactionism*, p. 25.
6. David B. Downing and Susan Bazargan, "Image and Ideology: Some Preliminary Histories and Polemics," in Downing and Bazargan, eds., *Images and Ideology in Modern/Postmodern Discourse*, p. 3.
7. For an introduction to the material culture literature, see G. S. Dunbar, "Illustrations of the American Earth: A Bibliographical Essay on the Cultural Geography of the United States," in *American Studies: An International Newsletter*, supplement to *American Quarterly* 25 (Autumn 1973), pp. 3–15; Peirce F. Lewis, "Learning from Looking: Geographic and Other Writing about the American Landscape," *American Quarterly* 35 (1983), pp. 242–61; Lewis, "Common Landscapes As Historic Documents," in Steven Lubar and W. David Kingery, eds., *History from Things: Essays on Material Culture*; and Thomas J. Schlereth, *Cultural History and Material Culture: Everyday Life, Landscapes, Museums*.
8. John A. Jakle and Keith A. Sculle, *The Gas Station in America*; John A. Jakle, Keith A. Sculle, and Jefferson S. Rogers, *The Motel in America*; John A. Jakle and Keith A. Sculle, *Fast Food: Roadside Restaurants in the Automobile Age*.
9. For example, see O. F. G. Sitwell, S. E. Olena, and S. E. Bilash, "Analyzing the Cultural Landscape As a Means of Probing the Non-Material Dimension of Reality," *Canadian Geographer* 30 (1986), pp. 132–45; and M. Gottdiener, *Postmodern Semiotics: Material Culture and the Forms of Postmodern Life*.
10. Keith L. Alexander, "Billboards Help Media Firms Weather Slowdown," *USA Today*, Dec. 12, 2000, p. 6B.
11. Roberta Henderson, "The Enduring Curse of Billboard Blight," *Louisville Courier Journal*, Nov. 2, 1997, p. D-3.
12. Alexander, "Billboards Help."
13. *Websters New International Dictionary of the English Language*, 2nd ed., s.v. "sign."
14. Ibid., s.v. "signboard."
15. James Claus, R. M. Oliphant, and Karen Claus, *Signs: Legal Rights and Aesthetic Considerations*, p. 167.
16. Karen E. Claus and R. James Claus, *The Sign Users Guide: A Marketing Aid*, p. 5.

1. SIGNS DOWNTOWN

1. See Paul C. Adams, "Peripatetic Imagery and Peripatetic Sense of Place," in Paul C. Adams, Steven Hoelscher, and Karen E. Till, eds., *Textures of Place: Exploring Humanist Geographies*, pp. 186–206.
2. George Williams, "Electricity for National Advertising," *Signs of the Times* 9 (Aug. 1909), p. 3.
3. Bury Irwin Dascent, "Electricity As Applied to Commercial Advertising," *Journal of Electricity, Power and Gas* 16 (Feb. 1906), pp. 99–101.
4. James H. Betts, "Flashers Make Electrics Compelling and Lengthen Life of Lamps," *Signs of the Times* 59 (June 1925), p. 52.
5. See C. A. Atherton, "Electric Signs of the Future Must Make Sales: Their 'Good Will' Days Are Over," *Signs of the Times* 42 (Aug. 1922), pp. 45–53; (Sept. 1922), pp. 46–51.
6. For a fuller discussion of electric storefront signs, see John A. Jakle, *City Lights: Illuminating the American Night*, pp. 211–23.
7. Philip Tocker, "Standardized Outdoor Advertising: History, Economics, and Self-Regulation," in John W. Houck, ed., *Outdoor Advertising: Its History and Regulation*, pp. 24–25.
8. For a brief review of the historical conditions that fostered consumer culture in the United States, see Michael Schudson, *Advertising, the Uneasy Persuasion: Its Dubious Impact on American Society*, pp. 147–61.
9. William Leiss, Stephen Kline, and Sut Jhally, *Social Communication in Advertising: Persons, Products, and Images of Well-Being*, 2nd ed., p. 102; Stuart Ewen, *Captains of Consciousness: Advertising and the Social Roots of the Consumer Culture*, pp. 51–52; Susan Strasser, *Satisfaction Guaranteed: The Making of the American Mass Market*, pp. 17–18.
10. Strasser, *Satisfaction Guaranteed*, p. 91; Frank Presbrey, *The History and Development of Advertising*, p. 511.
11. Strasser, *Satisfaction Guaranteed*, pp. 91, 93, 102; Victor Margolin, Ira Brichta, and Vivian Brichta, *The Promise and the Product: 200 Years of American Advertising Posters*, p. 31; Stephen Fox, *The Mirror Makers: A History of American Advertising and Its Creators*, pp. 30–31, 79; James D. Norris, *Advertising and the Transformation of American Society, 1865–1920*, pp. 138–39.
12. For an introduction to semiotics as applied to the study of landscape, see Jon Lang, "Symbolic Aesthetics in Architecture: Toward a Research Agenda," in Jack L. Nassar, ed., *Environmental Aesthetics: Theory, Research, and Applications*, pp. 11–26; Sitwell, Olenka, and Bilash, "Analyzing the Cultural Landscape." Linguist Ferdinand de Saussure and philosopher Charles Sanders Peirce first elucidated the tripartite schema but took no pains to illustrate its various uses, as did those who subsequently addressed the epistemology of various academic fields. For a discussion, see Gottdiener, *Postmodern Semiotics*. Henry Glassie, in "Structure and Function, Folklore and Artifact," *Semiotica* 7 (1973), pp. 313–51, was first to propose the theory of signs for the study of material culture.
13. Leiss, Kline, and Jhally, *Social Communication*, p. 9.
14. Norris, *Advertising*, p. 168.
15. Leiss, Kline, and Jhally, *Social Communication*, p. 5.

16. See Robert Sack, *Place, Modernity, and the Consumer's World: A Relational Framework for Geographical Analysis.*
17. Wilmot Lippincott, *Outdoor Advertising*, pp. 23–25.
18. Burton Harrington, "What Is Poster Advertising?" *Poster* 14 (Oct. 1923), p. 3.
19. E. L. Francis, "Structural Standardization in Outdoor Advertising," *Advertising Outdoors* 19 (July 1928), pp. 7–10; "The Structures of Organized Outdoor Advertising," *Advertising Outdoors* 21 (Dec. 1931), pp. 26–30, 42; James Fraser, *The American Billboard: 100 Years*, p. 14; Tocker, "Standardized Outdoor Advertising," p. 34.
20. A. L. McCarthy, "Eureka Launches National Campaign," *Poster* 15 (Apr. 1924), pp. 18–19, 23.
21. Presbrey, *History and Development*, p. 500; Tocker, "Standardized Outdoor Advertising," pp. 26–27, 29; Fraser, *American Billboard*, pp. 10–11; [David H. Souder], "A Glorious Half Century of Outdoor Advertising," *Signs of the Times* 143 (May 1956); Margolin, Brichta, and Brichta, *Promise and Product*, pp. 35–36.
22. Fraser, *American Billboard*, p. 12; Tocker, "Standardized Outdoor Advertising," p. 31; Hugh E. Agnew, *Outdoor Advertising*, p. 12; Souder, "Glorious Half Century," p. 119; H. C. Menefee, "W. H. Donaldson, Founder and Former Owner of Signs of the Times Dead," *Signs of the Times* 51 (Sept. 1925), p. 8; E. Thomas Kelley, "Signs of the Times Celebrates Its Twentieth Year of Service," *Signs of the Times* 52 (May 1926), p. 6.
23. Bill Dorsey and Tod Swormstedt, "A History of the Sign Industry in America, 1906–1981," *Signs of the Times* 203 (May 1981), p. 10; Fraser, *American Billboard*, p. 68.
24. Souder, "Glorious Half Century," pp. 114, 118, 120, 128; Fraser, *American Billboard*, p. 48; Moody's Investors Service, *Moody's Manual of Investments and Securities*, pp. 1291–92; Foster and Kleiser, *Fifty Years of Outdoor Advertising, 1901–1951*, p. 2.
25. Dorsey and Swormstedt, "History of the Sign Industry," pp. 10–14; www.oaaa.org/zzroutside/PublSide/AboutMedium?history.html (accessed Aug. 9, 2001).
26. G. H. S. Young, "Electrical Advertising in Boston," *Signs of the Times* 19 (Nov. 1912), p. 24.
27. S. N. Holiday, "Through the Years with Electrical Advertising on the Great White Way," *Signs of the Times* 68 (May 1931), p. 57.
28. Arthur Williams, "Broadway — a Fascinating Electric Sign Gallery — America's Brightest and Busiest Street," *Signs of the Times* 35 (Mar. 1917), p. 5.
29. "The World's Largest Spectacular," *Signs of the Times* 32 (Mar. 1917), p. 9.
30. For a fuller discussion of Broadway and Times Square sign spectaculars, see Jakle, *City Lights*, pp. 195–209.

2. SIGNS ON MAIN STREET

1. Sinclair Lewis, *Main Street*, p. 38.
2. Leroy Jacks, "Electrical Advertising Opportunities in Small Cities," *Signs of the Times* 15 (Nov. 1, 1913), p. 10.

3. "Illumination Is the Symbol of Cheer and the Sign of Patriotism," *Signs of the Times*, Apr. 1918, p. 34.
4. Gary Cross, *An All-Consuming Century: Why Commercialism Won in Modern America*, p. 17.
5. Ibid., p. 18.
6. U.S. Department of Transportation, Federal Highway Administration, *Highway Statistics: Summary to 1985*, table MV-291.
7. See Richard Longstreth, "Compositional Types in American Commercial Architecture," in Camille Wells, ed., *Perspectives in Vernacular Architecture, II*, pp. 12–24.
8. See Richard Longstreth, *The Buildings of Main Street: A Guide to American Commercial Architecture.*
9. "Bids on Signs for Entire Store Front from Maine to California," *Signs of the Times* 50 (July 1925), p. 82.
10. See J. N. Halsted, "Modern Sign Making with Lacquer," *Signs of the Times* 60 (July 1930), pp. 24–26.
11. Chester H. Liebs, *Main Street to Miracle Mile*, p. 212.
12. For an introduction to the place-product-packaging concept and a discussion of its historical application in gasoline retailing, see Jakle and Sculle, *Gas Station in America.*
13. See "Setting the Course, 1900–1930," in Cross, *All-Consuming Century*, pp. 17–65.
14. "National Advertising at the Point of Sale," *Signs of the Times* 79 (Mar. 1935), p. 9.
15. "Record-Breaking Sign Order for Schlitz Beer," *Signs of the Times* 76 (Mar. 1934), p. 19.
16. C. M. Lemperly, "A Bulletin Campaign of Giant Trade-Mark Cutouts," *Signs of the Times* 73 (June 1935), p. 22; "Big One in the West," *Signs of the Times* 92 (June 1939), p. 82.
17. "Buick's Identification Program," *Signs of the Times* 86 (Aug. 1937), p. 20.
18. A. C. Townsend, "If You Are Feeling Effects of the Sign Rental Bogey . . . Why Lease?" *Signs of the Times* 74 (July 1933), p. 17.
19. "Advertising Investment in 1,100,00 Signs . . . More than $205,000,000 Yearly," *Signs of the Times* 84 (Dec. 1936), p. 7.
20. Anne Friedberg, *Window Shopping: Cinema and the Postmodern*, p. 57.
21. Leonard S. Marcus, *The American Store Window*, p. 18.
22. See Richard Mattson, "Store Front Remodeling on Main Street," *Journal of Cultural Geography* 3 (Spring/Summer 1983), pp. 41–55.
23. Francis M. Fulge, "Architectural Sign Lighting Developments," *Signs of the Times* 71 (May 1932), pp. 17–18.
24. Mattson, "Store Front Remodeling," p. 49.
25. Lynn B. Blaine, "The Dawning of Awnings," *Signs of the Times* 184 (Aug. 1984), pp. 41–43.
26. Andrew Bertucci, "Electric Awning Signs Legislation and Zoning," *Signs of the Times* 223 (Jan. 2001), pp. 106–7.

27. Susan Conner, "2000 Digital Service Bureau Survey," *Signs of the Times* 223 (Jan. 2001), pp. 78–80.

3. ROADSIDE SIGNS

1. Quinta Scott and Susan Croce Kelly, *Route 66: The Highway and Its People*, p. 3; Ulrich Keller, *The Highway As Habitat: A Roy Stryker Documentation, 1943–1955*, pp. 15–19.
2. Thomas W. Salmon II, "Messages along the Roadside Landscape: A Vanishing Art Remembered with Nostalgia," *APT Bulletin* 30, no. 1 (1999), pp. 62–63.
3. Schudson, *Advertising*, 147–61; Fox, *Mirror Makers*, pp. 53–54.
4. Cross, *All-Consuming Century*, pp. 2–5.
5. Souder, "Glorious Half Century," pp. 113–14; Fraser, *American Billboard*, p. 68; Agnew, *Outdoor Advertising*, pp. 18–19; Dorsey and Swormstedt, "History of the Sign Industry," p. 15.
6. "Quantity Signs: Progressive Years," *Signs of the Times* 143 (May 1956), p. 163. Alert observers remarked about the considerable proliferation of signs along city streets in the decade after the close of the Civil War; see David E. Shi, *Facing Facts: Realism in American Thought and Culture, 1850–1920*, pp. 93–94.
7. Lippincott, *Outdoor Advertising*, p. 107; "Largest Service Station Sign Erected by Halfer and Ragg," *Signs of the Times* 52 (Mar. 1926), p. 54.
8. Kerwin H. Fulton, "From Fencepost Daubs to De Luxe Showings Is Advance Made in Poster Advertising," *Signs of the Times* 41 (July 1922), p. 6; Outdoor Advertising Association of America, *Outdoor Advertising — the Modern Marketing Force*, pp. 210, 218–19, 222.
9. "Studebaker Uses Signs Profusely — And Why," *Signs of the Times* 28 (Mar. 1915), p. 3.
10. "Outdoor Advertising for Automobiles," *Signs of the Times* 33 (Sept. 1, 1913), p. 3.
11. Fulton, "From Fencepost Daubs," p. 6.
12. M. Igert, "The Psychological Laws of Posters, Part I," *Poster* 18 (Mar. 1927), p. 18.
13. R. Fayerweather Babcock, "Education Department," *Signs of the Times* 18 (May 1927), pp. 21–22.
14. "A Model Bulletin Sign," *Signs of the Times* 9 (Nov. 1909), p. 24.
15. Sam Bass Warner, Jr., *Streetcar Suburbs: The Process of Growth in Boston, 1870–1900*; Liebs, *Main Street to Miracle Mile*.
16. Railway Station Advertising, Advertisement, *Signs of the Times* 14 (Feb. 1911), p. 19; Lippincott, *Outdoor Advertising*, p. 105.
17. James J. Flink, *The Automobile Age*, pp. 170–71.
18. Foster and Kleiser, Advertisement, *Poster* 14 (Apr. 1923), p. 3.
19. E. Thomas Kelley, "Changing Habits of Americans Rapidly Increasing Outdoor Circulation," *Signs of the Times* 47 (June 1927), p. 11; Fulton, "From Fencepost Daubs," p. 7; "Double Service Bulletins 'Guarding' Down East Roads Get Publicity for Hood Tires," *Signs of the Times* 37 (Oct. 1917), p. 26;

"Road Bulletins Increase Sales Over Seven Times for Hood Tires in New England," *Signs of the Times* 28 (Apr. 1918), p. 44.

20. Noah Young, Greyhound bus driver, 1955–1985, interview with Keith A. Sculle, Indianapolis, Ind., May 31, 2001; Frank Rowsome, Jr., *The Verse by the Side of the Road: The Story of the Burma-Shave Signs and Jingles.*
21. Outdoor Advertising Association of America, *Outdoor Advertising*, pp. 219, 222.
22. Fraser, *American Billboard*, pp. 14, 53, 102; Souder, "Glorious Half Century," pp. 114, 145; Margolin, Brichta, and Brichta, *Promise and the Product*, p. 31; Fox, *Mirror Makers*, pp. 30–31, 79; Presbrey, *History and Development of Advertising*, p. 511; Norris, *Advertising and the Transformation*, pp. 138–39.
23. A. V. Foley, "Experience of the Tide Water Oil Company in the Use of Outdoor Advertising," *Signs of the Times* 39 (Nov. 1921), pp. 9–10, 74. "Almost a Million," *Signs of the Times* 76 (Apr. 1931), p. 67; "Quality Signs," p. 163; "Efficient, Economical," *GE Review* 42 (Mar. 1939), p. 225.
24. Felicia Feaster and Bret Wood, *Forbidden Fruit: The Golden Age of the Exploitation Film*, pp. 8–11, 97–99.
25. Kenneth L. Roberts, "Travels in Billboardia," *Saturday Evening Post*, Oct. 13, 1928, pp. 24–25, 186, 189–90.
26. Jim Heimann and Rip Georges, *California Crazy: Roadside Vernacular Architecture*; Jim Heimann, *California Crazy and Beyond*; Karal Ann Marling, *Colossus of Roads: Myth and Symbol along the American Highway*, pp. 2–3.
27. Robert Venturi, Denise Scott Brown, and Steven Izenour, *Learning from Las Vegas.*
28. "Super Markets Step Up Their Use of Advertising Displays," *Signs of the Times* 137 (July 1954), pp. 23–26.
29. J. A. Prince, "Speed in Conveying Message Becomes More Vital Each Year," *Signs of the Times* 147 (Oct. 1957), pp. 42–43; "Signman's Ingenuity Creates a Spectacular," *Signs of the Times* 133 (Feb. 1933), pp. 23–24; "Chrysler Offers All Dealers Uniform Used Car Signs," *Signs of the Times* 151 (Jan. 1959), pp. 31, 101.
30. Alan Hess, *Viva Las Vegas: After-Hours Architecture*, pp. 118–19, 123.
31. Richard V. Francaviglia, *Main Street Revisited: Time, Space, and Image Building in Small-Town America*, pp. 57–59.
32. Dorsey and Swormstedt, "History of the Sign Industry," p. 32.
33. Claus, Oliphant, and Claus, *Signs*, pp. 66–67.
34. "Proposed Laws Threaten Rural Property Owners," *Signs of the Times* 131 (Aug. 1952), pp. 40–41.
35. Dorsey and Swormstedt, "History of the Sign Industry," p. 30.
36. Fraser, *American Billboard*, p. 188; Claus, Oliphant, and Claus, *Signs*, pp. 66–69.
37. Barry Meier, "Lost Horizons: The Billboard Prepares to Give Up Smoking," *New York Times*, Apr. 19, 1999, pp. A1, A20.
38. Alexander, "Billboards Help Media Firms," p. 6B. For a synopsis of outdoor advertising's virtues relative to other advertising sign types and testimony to the considerable extent to which that rationale convinced the business world, see the observations in the authoritative Standard and Poor's Corp.,

Standard Corporation Descriptions 59, no. 22 (Nov. 25, 1988), sec. 2, p. 3231, col. 2.

39. Moody's Investors Service, *Moody's Industrial Manual* (1963), pp. 2924–26; (1965), pp. 1368–69; "GOA, F & K Contend for No. 1 Rank in Outdoor Field; Naegle Moves Up," *Advertising Age* 34 (July 1, 1963), p. 79; Rich Zuhradnik, "Outdoor Obsession," *Marketing and Media Decisions* 21, no. 3 (Mar. 1986), p. 4; Moody's Investors Service, *Industrial Manual* (1989), p. 2947; Standard and Poor's, *Standard Corporation Descriptions*, pp. 3231–32; Moody's Investors Service, *Industrial Manual* (2000), p. 6266; "Lamar Outdoor Advertising Company," www.lamar.com/main/about/History.cfm (accessed Oct. 10, 2001).
40. Dorsey and Swormstedt, "History of the Sign Industry," pp. 11, 18; Fraser, *American Billboard*, p. 96.
41. Fraser, *American Billboard*, p. 99, 148.
42. Souder, "Glorious Half Century," p. 153; Dorsey and Swormstedt, "History of the Sign Industry," p. 22; Fraser, *American Billboard*, p. 99, 102.
43. Dorsey and Swormstedt, "History of the Sign Industry," pp. 24, 31; Claus, Oliphant, and Claus, *Signs*, pp. 35, 69; Jennifer Flinchpaugh and Linda Kitchen, "Flex in the City (and Everywhere Else)," *Signs of the Times* 222 (Sept. 2000), p. 106; Justin Green, "Sign Game," *Signs of the Times* 223 (Apr. 2001), p. 9.

4. TRAFFIC SIGNS

1. Gordon M. Sessions, *Traffic Devices: Historical Aspects Thereof*. The title of E. A. Mueller's "Aspects of the History of Traffic Signals," *IEE Transactions on Vehicular Technology* 19, no. 1 (Feb. 1970), pp. 6–17, accurately suggests the author's partial treatment of the subject, which is concerned exclusively with the devices, not their broader social context, and halts at the start of the era of standardization in 1935.
2. Boston Redevelopment Authority and U.S. Department of Housing and Urban Development, *City Signs and Lights: A Policy Study*, p. 112.
3. Bruce Seely, *Building the American Highway System: Engineers as Policy Makers*, pp. 64–65, comments on the persuasive aura of scientific objectivity. For example, an article about Connecticut's highway development, "Road Development Does Outstanding Beautification Work," *Roadside Bulletin* 2, no. 5 (Jan. 1934), pp. 1, 8, demonstrates how Elizabeth Boyd Lawton, foremost spokesperson in the 1930s for roadside beautification, deferred to public authority over traffic signs, although questioning cases of its insensitive landscaping.
4. T. W. Forbes, Thomas E. Snyder, and Richard F. Pain, "Traffic Sign Requirements," in Highway Research Board, *Night Visibility 1963 and 1964*, Highway Research Record 70, p. 48. Early in his discourse about how to advance the "safety movement," Sidney J. Williams, "Finding the Causes of Accidents," *Annals of the American Academy of Political and Social Science* 133 (Sept. 1927), pp. 156–60, emphasizes the complexity of driving.
5. John A. Montgomery, *Eno: The Man and the Foundation: A Chronicle of Transportation*, pp. 17, 33, 36, 40, 63.

6. James J. Flink, *America Adopts the Automobile, 1895–1910*, pp. 34–35; Clay McShane, *Down the Asphalt Path: The Automobile and the American City*, pp. 114–15; "The Dangers of Traffic: The Psychology of Accidents," *Scientific American*, supplement no. 1755 (Aug. 21, 1909), p. 127.
7. Robert L. Meyer, "A Look at the First Congress," *Safety and Health* 135, no. 2 (Feb. 1987), p. 23; Meyer, "The Council between the Wars," *Safety and Health* 135, no. 5 (May 1987), p. 22; Seely, *Building the American Highway System*, pp. 41–42; Sessions, *Traffic Devices*, pp. 85–86; Christy Borth, *Mankind on the Move: The Story of Highways*, pp. 210–11. For a history of traffic accidents at railroad crossings, see John R. Stilgoe, *Metropolitan Corridor: Railroads and the American Scene*, pp. 167–88.
8. For example, see "Highway Safety Signs," *Signs of the Times* 85 (Jan. 1937), p. 96, regarding Metropolitan Life Insurance Company's advertisements in 1937. For an example of the concern of private citizenry to combat traffic fatalities, see "Educating Drivers," *Public Safety* 5 (Sept. 1931), pp. 6–7.
9. Arthur F. Loewe, "Selling Safety on the Highways," *Signs of the Times* 85 (Mar. 1937), p. 46.
10. Nelson S. Spencer, "Street Signs and Fixtures," *Municipal Affairs* 5 (Sept. 1901), p. 726.
11. "National Legislation Proposed for Road Signs — Congress Asked to Pass Protection Law," *Signs of the Times* 32 (Nov. 1919), p. 65.
12. M. G. Lloyd, "Uniform Traffic Signs, Signals, and Markings," *Annals of the American Academy of Political and Social Science* 133 (Sept. 1927), p. 121.
13. Drake Hokanson, *The Lincoln Highway: Main Street across America*, pp. 17–18; John A. Jakle, "Pioneer Roads: America's Early Twentieth-Century Named Highways," *Material Culture* 32, no. 2 (2000), pp. 2–3; "Diminishing Traffic Law Fractures with the Aid of the Bulletin," *Signs of the Times* (Aug. 1923), p. 21.
14. "Traffic Signs Offer Opportunities to Sign Men — How They Aid in Making Service Complete," *Signs of the Times* 37 (Oct. 1917), p. 38.
15. Meyer, "Look at the First Congress" (Feb. 1987), p. 22; Ellis L. Armstrong, Michael C. Robinson, and Suellen M. Hoy, eds., *History of Public Works in the United States, 1776–1976*, pp. 124–25; Harry E. Neal, "Proper Marking of Highways Essential to Attain Maximum Services," *Signs of the Times* 51 (Dec. 1925), p. 6; "At Last — Uniform Traffic Signs," *Public Safety* 3 (Mar. 1929); Sessions, *Traffic Devices*, pp. 89, 117–20; Hokanson, *Lincoln Highway*, pp. 18, 106, 109.
16. U.S. Code Congressional Service, *Laws of Congress* 78th Cong., 2nd sess., p. 1349; Title 23, United States Code, Sect. 109(d); U.S. Department of Transportation, Federal Highway Administration, *Manual on Uniform Control Devices (MUTCD)*, mutcd.fhwa.dot.gov/kno-faq.htm (accessed Feb. 2, 2002); Armstrong, *History of Public Works*, p. 125.
17. J. M. Bennett, *Roadsides: The Front Yard of the Nation*, p. 161.
18. George W. Howie, "Traffic Signaling," pp. 350–81 in John E. Baerwald, ed., *Traffic Engineering Handbook*; C. Paul Jones and Eugene M. Wilson, "Advanced Signing for Recreational and Historic Sites," *Traffic Control Devices for Highways, Work Zones, and Railroad Grade Crossings*, p. 8.

19. Le Corbusier, *When the Cathedrals Were White*, p. 69; also see Galway Kinnell, "The Avenue Bearing the Initial of Christ into the New World" in Mike Marqusee and Bill Harris, eds., *New York: An Anthology*, p. 93.
20. For example, see Phil Patton, *Open Road: A Celebration of the American Highway*, pp. 132–33; Michael Wallis, *Route 66: The Mother Road*, pp. 26–27.
21. Armstrong, Robinson, and Hoy, *History of Public Works*, p. 123; Charles W. Stark, "New York's Early Traffic Signal Towers," *Traffic Engineering* 39 (Nov. 1968), pp. 48–49; Edward A. Mueller, "The Transportation Profession in the Bicentennial Year, Part II," *Traffic Engineering* 46 (Sept. 1976), p. 31; Frederick C. Gamst, "Rethinking Leach's Structural Analysis of Color and Instructional Categories in Traffic Control Signals," *American Ethnologist* 2, no. 2 (May 1975), p. 278; W. T. Perry, "A New Way to Control Traffic on Congested City Streets," *American City* 22 (1920), p. 476.
22. "A Traffic Light That Stays in Place," *American City* 24 (Apr. 1921), p. 443.
23. Borth, *Mankind on the Move*, p. 204; "Traffic Light That Stays in Place," p. 443; Stark, "New York's Early," p. 49; Gamst, "Rethinking Leach's Structural Analysis," p. 278; Sessions, *Traffic Devices*, p. 63.
24. Armstrong, Robinson, and Hoy, *History of Public Works*, p. 123; M. G. Lay, *Ways of the World: A History of the World's Roads and of the Vehicles That Used Them*, p. 187; Gamst, "Rethinking Leach's Structural Analysis," p. 281; Mueller, "Transportation Profession," p. 33; Sessions, *Traffic Devices*, pp. 43, 56; Armstrong, Robinson, and Hoy, *History of Public Works*, pp. 123–24; Stark, "New York's Early," p. 49.
25. Lloyd, "Uniform Traffic Signals," p. 122.
26. Arroway Traffic Signal Corp., Advertisement, *Public Safety* 10 (May 1936), p. 33.
27. "Traffic Experts Analyze 'Through Streets,'" *Public Safety* 4 (Jan. 1930), pp. 7–8.
28. Donald S. Berry, "They Shall Not Pass!" *Public Safety* 14 (Apr. 1939), pp. 20–21, 38.
29. Sessions, *Traffic Devices*, p. 134; for an example, see Eagle Sign Corporation, advertisement, *Public Safety* 9 (May 1935), p. 27.
30. "Effective Types of Danger Signals," *Concrete Highway Association* 2 (Mar. 1918), p. 93; George R. Chatburn, *Highways and Highway Transportation*, pp. 436, 439–40; "How California Marks Its Highways — 60,000 Direction Signs Placed," *Signs of the Times* 31 (Mar. 1919), p. 44; "3,500 Signs Placed in the West by Auto Club of Southern Cal." *Signs of the Times* 48 (Nov. 1924), p. 66. For an overview of the Automobile Club of Southern California's signposting program, beginning in 1905 and continuing through the end of the century, see Kristin Tilford, "Sign of the Times," *Westways* 91, no. 4 (July/Aug. 1999), pp. 26–30.
31. Sessions, *Traffic Devices*, p. 126.
32. "Reflecting Warning Signs," *American City* 45 (Mar. 1931), p. 173.
33. Highway Lighthouses, Advertistment, *Public Safety* 43 (June 1931), p. 43.
34. W. A. Van Duzer, "Don't Walk! Walk!" *Public Safety* 20 (July 1941), pp. 20–21.
35. "Slogans and Rhymes on Signs Help Keep Motorists Alert," *Signs of the Times* 145 (Feb. 1957), p. 41.

36. C. C. Wiley, “Road Signs for Night Driving,” *Roads and Streets* 76 (Feb. 1933), pp. 75–78; Sessions, *Traffic Devices*, pp. 124–25; “Electronic Signs Aid Traffic Control,” *Signs of the Times* 196 (Jan. 1974), p. 46.
37. Armstrong, Robinson, and Hoy, *History of Public Works*, pp. 127–28.
38. “New York’s New Street Signs,” *Electrical Age* 27 (July 1902), p. 272.
39. Armstrong, Robinson, and Hoy, *History of Public Works*, p. 124; Cliff Henderson, “Route Signs . . . Yesterday and Today,” *Signs of the Times* 148 (Jan. 1958), pp. 74, 100.
40. Chatburn, *Highways*, p. 462.
41. A. R. Hirst, “How Wisconsin Marks Its Highways — 5,000 Mile System Now Signed,” *Signs of the Times* 31 (Feb. 1919), p. 44.
42. “North Carolina’s Road Signs — ‘Clock System’ Used,” *Signs of the Times* 31 (Apr. 1919), p. 45.
43. “New Road Signs to Make Ohio Roads Best Marked in Country,” *Signs of the Times* 44 (July 1923), p. 61. Also see “Systematic Marking of Wayne County Highways,” *Concrete Highway Magazine* 2 (Oct. 1918), p. 239.
44. “Connecticut Color Route Sign System — Operation of the ‘Eastern Plan,’” *Signs of the Times* 31 (Feb. 1919), p. 45.
45. W. H. Jersig, “San Antonio’s Miniature Bulletin Boards Are Dedicated to the Weary Traveler,” *Signs of the Times* 15 (Dec. 1923), p. 12; Carroll Carson Wiley, *Principles of Highway Engineering*, p. 477; Harold A. Meeks, *On the Road to Yellowstone: The Yellowstone Trail and American Highways, 1900–1930*, pp. 91, 168.
46. For example, see “New Type of Reflecting Highway Signs,” *Signs of the Times* 132 (Oct. 1952), pp. 42, 139.
47. American Plywood Association, Advertisement, *Signs of the Times* 167 (May 1964), pp. 20–21.
48. Hayes E. Ross and Lawrence J. Sweeney, “Evaluation of Bolted-Base Steel Channel Signpost,” *Roadside Safety Appurtenances*, pp. 41–45; J. C. Powers, W. M. Szalaj, and R. L. Hollinger, “Breakaway Sign Testing, Phase I,” *Roadside Safety Appurtenances*, pp. 38–40; Ron E. Pelkey, “New Design Approach to Long-Span Overhead Sign Structures,” *Design of Sign Supports and Structures*, pp. 1–10; Kay Danielson, “Stately Signs,” *Signs of the Times* 214 (Feb. 1992), pp. 92–95; Sessions, *Traffic Devices*, p. 123.
49. Jones and Wilson, “Advanced Signing,” pp. 8–13; John Tymoski, “Park Service Goes Plastic,” *Signs of the Times* 217 (Nov. 1995), pp. 132–35.
50. Mary Beth Lewis, “Ten Best Road Signs,” *Car and Driver* 35, no. 7 (Jan. 1990), p. 118.

5. SIGNS AND COMMUNITY

1. Eileen Leonard, Herman Strasser, and Kenneth Westhues, eds., *In Search of Community: Essays in Memory of Werner Stark, 1909–1985*, p. 7; D. W. Meinig, “Symbolic Landscapes: Models of American Community,” in D. W. Meinig, ed., *The Interpretation of Ordinary Landscapes*, p. 174.
2. Neil Leach, “Siegfried Kracauer,” in Neil Leach, ed., *Rethinking Architecture: A Reader in Cultural Theory*, p. 51.
3. John P. Hewitt, *Dilemmas of the American Self*, pp. 12–13.

4. Robert Booth Fowler, *The Dance with Community: The Contemporary Debate in American Political Thought*, pp. 3–4.
5. Ibid., p. 3.
6. For example, see Philip Selznick, "In Search of Community," in William Vitek and Wes Jackson, eds., *Rooted in the Land: Essays on Community and Place*, p. 196.
7. Fowler, *Dance with Community*, p. 14.
8. Frank W. Young, "Review Essay: Putnam's Challenge to Community Sociology," *Rural Sociology* 66, no. 3 (2001), p. 468.
9. Fowler, *Dance with Community*, pp. 34–37.
10. Ibid., p. 36.
11. Robert Wuthnow, *Sharing the Journey: Support Groups and America's New Quest for Community*, p. 365.
12. William M. Sullivan, "A Public Philosophy for Civic Culture," in Vitek and Jackson, *Rooted in the Land*, p. 243; Wilbur Zelinsky, *Nation into State: The Shifting Symbolic Foundations of American Nationalism*, p. 203. For an example of partisanship for the Confederate flag, see Patrick J. Buchanan, *The Death of the West: How Dying Populations and Immigrant Invasions Imperil Our Country and Civilization*, pp. 259–60.
13. For example, see Charley Reese, "Flag Should Be Treated As Something Extra Special," *Danville (IL) Commercial News*, Nov. 23, 2001, p. 6A.
14. Olivia Barker, "'Star and Stripes' Is Making a Lone Star Statement," *USA Today*, July 6, 2001, p. D1.
15. "Forward America," *Signs of the Times* 64 (Jan. 1930), p. 94.
16. "O.A.A. Votes to Back Manufacturers in Campaign of Industrial Leaders to Promote Faith in 'the American Way,'" *Signs of the Times* 84 (Nov. 1936), p. 9.
17. Claude S. Fischer, "Ambivalent Communities: How Americans Understand Their Localities," in Alan Wolfe, ed., *America at Century's End*, p. 89.
18. Scott Russell Sanders, "The Common Life," in Vitek and Jackson, *Rooted in the Land*, p. 44.
19. James C. Klotter and John W. Muir, "Boss Ben Johnson, the Highway Commission, and Kentucky Politics, 1927–1937," *Register of the Kentucky Historical Society* 84, no. 1 (Winter 1986), pp. 37–38.
20. Harold J. Ashe, "How Painted Display Boosts San Diego," *Poster* 19 (Jan. 1928), p. 20.
21. Sessions, *Traffic Devices*, p. 94. For a summary of Fisher's work with some its material culture effects, see Hap Hatton, *Tropical Splendor: An Architectural History of Florida*, pp. 56–59. The full-scale biography, although with far more attention to the inner workings of Fisher's business than to its landscape results, is Mark S. Foster, *Castles in the Sand: The Life and Times of Carl Graham Fisher*.
22. Tilford, "Signs of the Times," p. 29.
23. Orville McPherson, "San Diego Welcomes Tourists with Bulletins along the Highways," *Signs of the Times* 58 (Apr. 1928), p. 12.
24. Ashe, "How Painted Sign Display," p. 20.
25. "Do the Road Signs Leading to Your City Advertise It Properly?" *Signs of the Times* 29 (Oct. 1915), p. 36.

26. Ibid.
27. "Signs Show the Way," *Lincoln Highway Journal* 2, no. 1 (Summer 1997), p. 3.
28. Olga Herbert, "Where the Rubber Meets the Road," *Lincoln Highway Journal* 2, no. 1 (Summer 1997), p. 2.
29. "'97 Lincoln Highway Road Rally," *Lincoln Highway Journal* 2: 2 (Fall 1997), p. 4.
30. Tom Teague, *Searching for 66*, p. 238. Marguerite S. Shaffer (*See America First: Tourism and National Identity, 1880–1940*, pp. 242–60) observed the search for self-discovery through tourism from the start of automobile tourism.
31. J. B. Jackson, "The Future of the Vernacular," in Paul Groth and Todd W. Bressi, eds., *Understanding Ordinary Landscapes*, p. 154.
32. For example, see Alan Ehrenhalt, "The Empty Square," *Preservation* 52, no. 2 (Mar./Apr. 2000), pp. 43–46, 48–51.
33. Hugh Russell Wilks, "Detroit Insurance Company Becomes Famous through Humorous Electrical Advertising," *Signs of the Times* 48 (Dec. 1924), p. 31.
34. Lilas S. Bill, "Posters a Factor in Good Citizenship," *Poster* 14 (Dec. 1923), p. 6.
35. "Opportunity, Progress and Prosperity Symbolized by the Slogan Sign — a City's Best Ad," *Signs of the Times* 35 (Feb. 1917), pp. 24–25.
36. For example, see "Cities That Have Slogan Signs," *Signs of the Times* (Nov. 1, 1913), p. 34; "Cities That Advertise by Means of Slogan Signs," *Signs of the Times* 23 (Dec. 1, 1913), p. 8.
37. "Opportunity, Progress and Prosperity," p. 24.
38. Lynn Simross, "Man behind the Hollywood Sign," *Los Angeles Times*, Jan. 18, 1977, part 4, pp. 1, 6.
39. "$10,000 Donation Saves Landmark Hollywood Sign," *Los Angeles Times*, Mar. 14, 1973, part 2, p. 1; Lynn Simross, "A Hollywood Cliff-Hanger," *Los Angeles Times*, Jan. 12, 1977, pp. 1, 7.
40. Wilbur Zelinsky, "Welcoming Signs along America's Highways: Where Every Town Is above Average," in *Exploring the Beloved Country: Geographic Forays into American Society and Culture*, p. 266.
41. Richard V. Francaviglia, *The Shape of Texas: Maps As Metaphors*, pp. 77–78.
42. Barbara A. Weightman, "Sign Geography," *Journal of Cultural Geography* 9, no. 1 (Fall/Winter 1988), p. 62.
43. Thomas A. Wikle, "Those Benevolent Boosters: Spatial Patterns of Kiwanis Membership," *Journal of Cultural Geography* 17, no. 1 (1996), pp. 1, 6–7.
44. James J. Parsons, "Hillside Letters in the Western Landscape," *Landscape* 30, no. 1 (1988), pp. 15–23.
45. James F. Barker, Michael W. Fazio, and Mark Hildebrandt, *The Small Town As an Art Object*, p. 97.
46. Kent C. Ryden, *Mapping the Invisible Landscape: Folklore, Writing, and the Sense of Place*, pp. 1–18.
47. For example, see Thomas O. Grieves, "What Posters Did for a Church," *Poster* 15 (Jan. 1924), pp. 26–27; "What I Learned from a Poster Board," *Poster* 15 (May 1924), pp. 11–12.
48. James Davison Hunter and John Steadman Rice, "Unlikely Alliances: The Changing Contours of American Religious Faith," in Wolfe, *America at Century's End*, p. 331.

49. Judith Stacey, "Background toward the Postmodern Family: Reflections on Gender, Kinship, and Class in the Silicon Valley," in Wolfe, *America at Century's End*, p. 18.
50. Kathleen Gerson, "Coping with Commitment: Dilemmas and Conflict of Family Life," in Wolfe, *America at Century's End*, p. 35.
51. Roy F. Baumeister, *Identity: Cultural Change and the Struggle for Self*, p. 72.

6. TERRITORIAL MARKERS AND SIGNS OF PERSONAL IDENTITY

1. Sut Jhally, *The Codes of Advertising: Fetishism and the Political Economy of Meaning in the Consumer Society*, p. 1.
2. Judith Williamson, *Decoding Advertisements: Ideology and Meaning in Advertising*, p. 43.
3. Diane Barthel, *Putting on Appearances: Gender and Advertising*, p. 3.
4. Guy Cook, *The Discourse of Advertising*, p. 177.
5. William M. O'Barr, *Culture and the Ad: Exploring Otherness in the World of Advertising*, p. 2.
6. Robert Goldman, *Reading Ads Socially*, p. 5.
7. Ibid., p. 18.
8. Leiss, Kline, and Jhally, *Social Communication in Advertising*, p. 285.
9. Ibid., p. 291.
10. Roland Marchand, *Advertising the American Dream: Making Way for Modernity, 1920–1940*, p. xvii.
11. E. M. Fairley, "Poster People," *Poster* 11 (Feb. 1920), p. 19.
12. Philip Chandler, "La Palina Cigars 'Made Good,'" *Poster* 16 (July 1925), p. 7.
13. Ibid.
14. Jhally, *Codes of Advertising*, p. 128.
15. David L. Altheide, "Identity and the Definition of the Situation in a Mass-Mediated Context," *Symbolic Interaction* 23, no. 1 (2000), p. 4.
16. Harold M. Proshansky, Abbe K. Fabian, and Robert Kaminoff, "Place-Identity: Physical World Socialization of the Self," *Journal of Environmental Psychology* 3 (1983), p. 58.
17. For a fuller discussion of place and personal identity, see David Kolb, "Self-Identity and Place," in *Postmodern Sophistications: Philosophy, Architecture, and Tradition*, pp. 146–58; and Clare L. Twigger-Ross and David L. Uzzell, "Place and Identity Processes," *Environmental Psychology* 16 (1996), pp. 205–20.
18. For a fuller discussion of social stereotyping, see Walter G. Stephan and David Rosenfield, "Racial and Ethnic Stereotypes," in Arthur G. Miller, ed., *In the Eye of the Beholder: Contemporary Issues in Stereotyping*, pp. 92–129; and Anthony J. Cortese, *Provocateur: Images of Women and Minorities in Advertising*.
19. See Judith Waters and George Ellis, "The Selling of Gender Identity," in Mary Cross, ed., *Advertising and Culture: Theoretical Perspectives*, pp. 91–103.
20. Charles Goodrum and Helen Dalrymple, *Advertising in America: The First 200 Years*, p. 249.
21. Wilma McKenzie, "Merchandising to Women," *Advertising Outdoors* 2 (June 1931), p. 15.
22. Evelyn C. Rucker, "Women and the Poster," *Poster* 17 (May 1926), p. 13.

23. Lucy Komisar, "The Image of Woman in Advertising," in Vivian Gornick and Barbara K. Moran, eds., *Woman in Sexist Society: Studies in Power and Powerlessness*, p. 306.
24. Barthel, *Putting on Appearances*, p. 171.
25. Komisar, "Image of Woman," p. 310.
26. Ibid.
27. See Marilyn Kern-Foxworth, *Aunt Jemima, Uncle Ben, and Rastus: Blacks in Advertising, Yesterday, Today, and Tomorrow*.
28. See Jannette L. Dates, "Advertising," in Jannette L. Dates and William Barlow, eds., *Split Image: African Americans in the Mass Media*, pp. 461–91.
29. Marchand, *Advertising the American Dream*, p. 360.
30. Cross, *All-Consuming Century*, p. 1.
31. Don Mitchell, *Cultural Geography: A Critical Introduction*, p. 67.
32. Cross, *All-Consuming Century*, p. 2.
33. Arnold Mitchell, *The Nine American Lifestyles*, pp. 3–24.
34. J. Douglas Porteous, *Environment and Behavior: Planning and Everyday Urban Life*, p. 25.
35. Robert David Sack, *Human Territoriality: Its Theory and History*, p. 19.
36. Robert Sommer, *Personal Space: The Behavioral Basis of Design*, p. 44.
37. For a discussion of body space and its relation to landscape, see Edward S. Casey, "Body, Self and Landscape: A Geographical Inquiry into the Place-World," in Paul C. Adams, Steven Hoelscher, and Karen E. Till, eds., *Textures of Place: Exploring Humanist Geographies*, pp. 403–25.
38. Erving Goffman, *The Presentation of Self in Everyday Life*, p. 4.
39. Charissa Jones, "Leaving Their Mark Behind," *USA Today*, Nov. 22, 2000, p. 19A
40. See David Ley and Roman Cybriwsky, "Urban Graffiti As Territorial Markers," *Annals, Association of American Geographers* 64 (Dec. 1974), pp. 491–505.
41. See Mark Francis, "Control As a Dimension of Public-Space Quality," in Irwin Altman and Ervin H. Zube, eds., *Public Places and Spaces*, pp. 147–71.
42. See Paul B. Harris and Barbara B. Brown, "The Home and Identity Display: Interpreting Resident Territoriality from Home Exteriors," *Environmental Psychology* 16 (1996), pp. 187–203.

7. SIGNS AND LANDSCAPE VISUALIZATION

1. For a similar conception of space, see Henri Lefebvre, *The Production of Space*, p. 33.
2. Gordon D. Logan and N. Jane Zbrodoff, "Selection for Cognition: Cognitive Constraints on Visual Spatial Attention," *Visual Cognition* 6 (1999), p. 61.
3. Ibid., p. 72.
4. David Canter, *The Psychology of Place*, p. 20.
5. J. M. Mandler, quoted in Giovanna Axia, Erminielda Mainardi Peron, and Marcia Rosa Baroni, "Environmental Assessment across the Life Span," in Tommy Gärling and Gary W. Evans, eds., *Environment, Cognition, and Action: An Integrated Approach*, p. 223.

6. Axia, Peron, and Baroni, "Environmental Assessment," p. 224.
7. John S. Pipkin, "Urban Geometry in Image and Discourse," in Paul L. Knox, ed., *The Design Professions and the Built Environment*, p. 70.
8. Ibid., p. 74.
9. Kenneth E. Boulding, *The Image: Knowledge in Life and Society*; Kevin Lynch, *The Image of the City*.
10. Canter, *Psychology of Place*, p. 21.
11. Lynch, *Image of the City*, pp. 46–48.
12. Donald Appleyard, "Why Buildings Are Known: A Predictive Tool for Architects and Planners," *Environment and Behavior* 1 (Dec. 1969), p. 136.
13. Reginald G. Golledge, "Cognition of Physical and Built Environment," in Gärling and Evans, *Environment, Cognition, and Action*, p. 45.
14. For an introduction to cognitive maps, see Roger M. Downs and David Stea, *Maps in Mind: Reflections on Cognitive Mapping*.
15. See Gordon Cullen, *The Concise Townscape*, p. 17.
16. James J. Gibson, *The Ecological Approach to Visual Perception*, p. 97.
17. Ian Nairn, *The American Landscape: A Critical View*, pp. 30–34.
18. Paul Arthur and Romedi Passini, *Wayfinding: People, Signs, and Architecture*, p. 33.
19. Jay Appleton, *The Experience of Landscape*, p. 70.
20. Christopher Tunnard and Boris Pushkarev, *Man-Made America: Chaos or Control?* p. 172.
21. Donald Appleyard, Kevin Lynch, and John R. Myer, *The View from the Road*, pp. 4–11.
22. See Stephen Carr and Dale Schissler, "The City As a Trip: Perceptual Selection and Memory in the View from the Road," *Environment and Behavior* 1 (June 1969), pp. 7–35; Russ Parsons et al., "The View from the Road: Implications for Stress Recovery and Immunization," *Environmental Psychology* 18 (1998), pp. 113–40.
23. Tunnard and Pushkarev, *Man-Made America*, p. 175.
24. Ibid., p. 177.
25. J. Todd Snow, "The New Road in the United States," *Landscape* 17 (Autumn 1967), p. 16.
26. E[dward] Relph, *Place and Placelessness*, pp. 87, 125.
27. Fred Fischer, "Ten Phases of the Animal Path: Behavior in Familiar Situations," in Aristide H. Esser, ed., *Behavior and Environment: The Use of Space by Animals and Men*, p. 13.
28. See Warren J. Belasco, *Americans on the Road: From Autocamp to Motel, 1910–1945*, p. 86.
29. James M. Flagg, *Boulevards All the Way — Maybe*, pp. 27–29.
30. Jones and Wilson, "Advanced Signing," p. 8.
31. See Jan Theeuwes, "Visual Selection: Exogenous and Endogenous Control," in A. D. Gale et al. eds., *Vision in Vehicles: III*, pp. 53–61.
32. Claus and Claus, *Sign Users Guide*, p. 95.
33. Ibid., p. 193.

34. Theresa J. B. Kline, Laura M. Ghali, and Donald W. Kline, "Visibility Distance of Highway Signs among Young, Middle-Aged, and Older Observers: Icons Are Better than Text," *Human Factors* 32 (Oct. 1990), p. 609.
35. Arthur and Passini, *Wayfinding*, p. 184.
36. William Saroyan, *Short Drive, Sweet Chariot*, p. 62.
37. Bennett, *Roadsides*, p. 10.
38. J. B. Priestley, *Midnight on the Desert*, pp. 86–88.
39. Ilya Ilf and Eugene Petrov, *Little Golden America*, p. 15.
40. Jeff Brouws, photographer, text by Bernd Polster and Phil Patton, *Highway: Americas Endless Dream*, p. 1.
41. Cook, *Discourse of Advertising*, p. 29.
42. John Robinson, *Highways and Our Environment*, p. 7.
43. Marchand, *Advertising the American Dream*, p. 152.
44. See Peirce F. Lewis, "The Geographer As Landscape Critic," in Peirce F. Lewis, David Lowenthal, and Yi-Fu Tuan, *Visual Blight in America*, p. 1.
45. David Lowenthal, "The Offended Eye: Toward an Excrescential Geography," in Lewis, Lowenthal, and Tuan, *Visual Blight in America*, pp. 29–31.
46. Harry Lucht, "Main Street, USA," *Roadside Bulletin* 2 (May 1933), p. 22.
47. C. S. Farnsworth, untitled verse, *Roadside Bulletin* 2, no. 3 (c. 1933), p. 16.
48. "Billboards Destroy Motoring Joy," *Roadside Journal* 2, No. 5 (Jan. 1934), p. 3.
49. "Advertising Litter Now Causing Real Barrier to Beauty," *Roadside Bulletin* 2 (May 1932), p. 4.
50. "Lurid Billboards Contribute to Annual Loss of Lives by Accidents on Highways," *Roadside Bulletin* 2 (May 1933), p. 1.
51. Pennsylvania Roadside Council, *This or That? (Scenery or Sign-ery?)*, p. 3.
52. A. C. Daschbach, "The Why of the Sign Board," *Signs of the Times* 27 (Oct. 1914), p. 6.
53. Harris Wescott, "If Standardized Outdoor Advertising Did Not Exist," *Poster* 19 (March 1928), p. 9.
54. James C. Wright, "Can the United States Have Both Beauty and Business?" *Signs of the Times* 177 (Dec. 1967), p. 40.
55. See Ross Barrett, "Fitting Outdoor Advertising into the Environment," *Signs of the Times* 197 (Oct. 1972), p. 32.
56. General Outdoor Advertising Co. *Annual Report for* 1957, p. 1.
57. Tom Martinson, *Signs of the City: The Role of Outdoor Advertising in Municipal Planning*, p. 21.
58. Richard Newman, "Censorship in America Today," in *Proceedings of the Urban Signage Forum*, p. 108.
59. John Ise, *Our National Park Policy: A Critical History*, p. 568.
60. Outdoor Advertising, Inc., *The Truth about Standardized Outdoor Advertising*, p. 53.

8. SIGN REGULATION

1. Richard W. Stevenson, "Challenging the Billboard Industry," *New York Times*, Aug. 30, 1988, p. 38.

2. Charles Mulford Robinson, *Modern Civic Art; or, The City Made Beautiful*, 3d ed., p. v.
3. William H. Wilson, "The Billboard: Bane of the City Beautiful," *Journal of Urban History* 13 (Aug. 1987), p. 400.
4. Regarding pastoralism, see Leo Marx, *The Machine in the Garden: Technology and the Pastoral Ideal in America*; James L. Machor, *Pastoral Cities: Urban Ideals and the Symbolic Landscape of America*.
5. William H. Wilson, *The City Beautiful Movement*, pp. 86–87.
6. John DeWitt Warner, "Advertising Run Mad," *Municipal Affairs* 4, no. 2 (June 1900), p. 277.
7. Wilson, "Billboard," pp. 400–401.
8. Warner, "Advertising Run Mad," pp. 279–82.
9. Wilson, *City Beautiful Movement*, p. 288.
10. Wilson, "Billboard," pp. 394–425. Not all City Beautiful reformers were successful in their cities, however; for example, see Kristin Szylvian Bailey, "Fighting 'Civic Smallpox': The Civic Club of Allegheny County's Campaign for Billboard Regulation, 1896–1917," *Western Pennsylvania Historical Magazine* 70, no. 1 (Jan. 1987), pp. 3–28.
11. Clinton Rodda, "The Accomplishment of Aesthetic Purposes under the Police Power," *Southern California Law Review* 27 (1954), p. 169; *Crawford v. Topeka*, 51 Kan. (1893) 756; "Victory for 'Sky Signs' in N.Y.," *Signs of the Times* 9 (May 1909), p. 7.
12. "Victory for 'Sky Signs' in N.Y.," p. 7.
13. *Churchill and Tait v. Rafferty*, 32 P.I. 38.
14. Ibid., p. 42.
15. Ibid., pp. 52–54.
16. Ibid., p. 58.
17. Rodda, "Accomplishment," p. 172.
18. Howard L. McBain, *American City Progress and the Law*, p. 90.
19. George Kriemm, "Chicago Made Hideous with Signboards," *Chicago Sunday Times-Herald*, May 6, 1900, p. 4-3.
20. *Crawford v. Topeka*, 51 Kan. (1893) 756; *Bill Posting Sign Company v. Atlantic City*, 58 Atl. (1904) 342; *City of Chicago v. Gunning System*, 73 N.E. (1905) 1035; *Commonwealth v. Boston Advertising Company*, 74 N.E. (1905) 601; *Passaic v. Patterson Bill Posting, Advertising and Sign Painting Co.*, 72 N.J.L. (1905) 287; *Varney and Green v. Williams (East San Jose)*, 100 Pac. (1909) 867; and *Curran Bill Posting and Distributing Company v. Denver*, 107 Pac. (1910) 261.
21. *Passaic v. Patterson Bill Posting, Advertising and Sign Painting Co.*, 72 N.J.L. (1905) 287.
22. Edwin C. Rafferty, "Orderly City, Orderly Lives: The City Beautiful Movement in St. Louis," *Gateway Heritage* 11, no. 4 (Spring 1991), pp. 40–62; Wilson, "Billboard," 417; *St. Louis Gunning Advertisement Co. v. City of St. Louis et al.*, 137 S.W. (1911) 932–33, 958; McBain, *American City Progress*, p. 80; Rodda, "Accomplishment," p. 169; Wilson, "Billboard," pp. 417–18; Newman F. Baker, "Municipal Aesthetics and the Law," *Illinois Law Review* 20 (Feb. 1926), p. 567, n. 115.

23. *City of Chicago et al. v. Gunning System*, 214 Ill. (1905) 642; *The Haller Sign Works v. The Physical Culture Training School*, 249 Ill. (1911) 442–43.
24. *The Thomas Cusack Company v. The City of Chicago*, 267 Ill. (1915) 345, 346.
25. McBain, *American City Progress*, pp. 85–86.
26. Robinson, *Modern Civic Art*, pp. 222, 333.
27. Chauncey Shafter Goodrich, "Billboard Regulation and the Aesthetic Viewpoint with Reference to California Highways," *California Law Review* 17 (1928–29), pp. 120–34.
28. Tom Nokes, "Helpful Suggestions for Cleaning Up Outdoor Displays," *Signs of the Times* 49 (Jan. 1925), p. 25; Souder, "Glorious Half-Century," p. 113; Presbrey, *History and Development*, pp. 565–66.
29. Dorsey and Swormstedt, "History of the Sign Industry," p. 10.
30. Presbrey, *History and Development*, p. 505.
31. Outdoor Advertising Association of America, *Outdoor Advertising*, pp. 203–4.
32. "Adverse Legislation — A Blessing in Disguise That Won Favor for Painted Display," *Signs of the Times* 44 (May 1923), p. 6; William J. King, "Meeting Destruction with Construction — Opposition without Opposition," *Signs of the Times* 44 (May 1923), pp. 8–9, 20–21; Interview with Kip Pope (owner, C & U Poster Co., 1984–2001), Aug. 28, 2001.
33. Daniel M. Bluestone, "Roadside Blight and the Reform of Commercial Architecture," in Jan Jennings, ed., *Roadside America: The Automobile in Design and Culture*, pp. 170–84.
34. Megg Maguire and Frank Vespe, "'I Think That I Shall Never See a Billboard Lovely As a Tree,'" *Forum Journal* 14, no. 4 (Summer 2000), pp. 57–58; Howard J. Curry, "New Maine Law Will Develop Outdoor Advertising to High Standard," *Signs of the Times* 51 (Sept. 1925), pp. 7, 14. Hawaii's experience is the only one of the states barring billboards that is currently documented; see Margit Misangyi Watts, *High Tea at Halekulani: Feminist Theory and American Clubwomen*, pp. 119–63.
35. *Massachusetts General Laws* (1932), p. 67.
36. George K. Gardner, "The Massachusetts Billboard Decision," *Harvard Law Review* 69, no. 6 (Apr. 1936), pp. 870–73.
37. "Proposed Statutes Reflecting Public Demand for Action," *Roadside Bulletin* 2, no. 1 (1932), pp. 1, 8; "Billboard Industry Seeks to Vitiate Bay State Statute," *Roadside Bulletin* 2, no. 4 (May 1933), pp. 18–19.
38. *General Outdoor Advertising Company, Inc., et al. v. Department of Public Works*, 289 Massachusetts (1935) 182, 169.
39. Gardner, "Massachusetts Billboard Decision," p. 889.
40. For example, see Presbrey, *History and Development*, pp. 608–18.
41. L. I. Hewes, "Control Roadside Advertising and Aesthetics," *Roads and Streets* 69 (Oct. 1929), p. 346.
42. "The Preservation of Scenic Highways," *Poster* (Dec. 1929), p. 21.
43. U. S. Webb to W. E. Stewart, Manager, Central Coast Division, California Development Association, Apr. 27, 1928, cited in "Preservation of Scenic Highways," p. 22.

44. "Zoning Plan Presented at Roadside Beauty Conference," *Advertising Outdoors* 2 (Apr. 1931), pp. 29, 31.
45. *Perlmutter Furniture Company et al. v. Greene, Superintendent of Public Works, et al.*, 182 N.E. (1932) 7.
46. *Kelbro, Inc., v. Myrick, Secretary of State, et al.*, 30 A. 2d (1943) 529.
47. *Perlmutter Furniture Company et al. v. Greene, Superintendent of Public Works, et al.*, 182 N.E. (1932) 6.
48. Michael Litka, "The Use of Eminent Domain and Police Power to Accomplish Aesthetic Goals," in John W. Houck, ed., *Outdoor Advertising: History and Regulation*, pp. 91–92; Rodda, "Accomplishment," p. 152.
49. "Senate Committee Eliminates Ban of Outdoor Signs from Road Bill," *Signs of the Times* 140 (June 1955), p. 27.
50. Clifton W. Enfield, "Federal Highway Beautification: Outdoor Advertising Control, Legislation and Regulation," in Houck, *Outdoor Advertising*, pp. 150–51.
51. Charles F. Floyd and Peter J. Shedd, *Highway Beautification: The Environmental Movement's Greatest Failure*, p. 65; "The Case for Highway Adv[ertising]," *Signs of the Times* 146 (May 1957), pp. 37, 107; Enfield, "Federal Highway Beautification," pp. 150–62; "Finalized Federal Standards Are Announced," *Signs of the Times* 150 (Dec. 1958), p. 52.
52. "Millions of Americans Depend on Highway Advertising," *Signs of the Times* 145 (Feb. 1957), p. 86.
53. Peter Blake, *God's Own Junkyard: The Planned Deterioration of America's Landscape*, pp. 11, 14.
54. Floyd and Shedd, *Highway Beautification*, p. 72; Charles Stevenson, "The Great Billboard Scandal of 1960," *Reader's Digest* 76 (Mar. 1960), pp. 146–56; "Kennedy Takes on the Billboard Barons," *Christian Century*, Apr. 12, 1961, p. 447.
55. Ruth I. Wilson, "Billboards and the Right to Be Seen from the Highway," *Georgetown Law Journal* 30 (1942), pp. 723–50.
56. Floyd and Shedd, *Highway Beautification*, pp. 72–73.
57. Philip Tocker, "Standardized Outdoor Advertising," pp. 53–54; Floyd and Shedd, *Highway Beautification*, pp. 78–81; American Society of Planning Officials, "The Elimination of Nonconforming Signs," Report no. 209 (Apr. 1966), p. 2. For a minute chronicle of the legislative maneuverings behind the Highway Beautification Act of 1965, see Lewis L. Gould, *Lady Bird Johnson and the Environment*, pp. 136–68.
58. Tocker, "Standardized Outdoor Advertising," pp. 54, 56; Floyd and Shedd, *Highway Beautification*, pp. 82–88.
59. "The Outdoors," *Time*, Mar. 29, 1968, p. 27; Maguire and Vespe, "'I Think That I Shall,'" p. 57; Christie Carter, "The Beckoning Archives: Tourism Records at the Vermont State Archives," *Vermont History* 65, no. 3/4 (Summer/Fall 1997), p. 170; Theodore M. Riehle, "No Billboards in Vermont," *American City* 83 (Sept. 1968), p. 166.
60. Floyd and Shedd, *Highway Beautification*, pp. 104, 106–10; Martin L. Bell and Richard Wendel, "State Compliance to HBA Will Be Costly," *Signs of the Times*

175 (Mar. 1967), p. 107; Ted Williams, "Litter on a Stick," *Audubon* 93 (Aug./Sept. 1991), p. 16; James Nathan Miller, "The Great Billboard Double-Cross," *Reader's Digest* 126 (July 1985), pp. 83–90.

61. Miller, "Great Billboard Double-Cross," pp. 89–90; National Council for the Protection of Roadside Beauty, www.scenic.org/national.htm (accessed Oct. 10, 2001).
62. Stevenson, "Challenging the Billboard Industry," pp. 1, 38.
63. Baker, "Municipal Aesthetics and the Law," pp. 571–72; Rodda, "Accomplishment," p. 150. For a list of the cases in the 1970s demonstrating the sole sufficiency of aesthetics for sign regulation, see Floyd and Shedd, *Highway Beautification*, pp. 60–61.
64. Floyd and Shedd, *Highway Beautification*, pp. 50–51.
65. *Metrommedia, Inc., et al. v. City of San Diego et al.*, 453 U.S. (1983) 501, 569.
66. Ibid., p. 505.
67. Ibid., pp. 497, 525.
68. Brad Sanders "The First Amendment 'Law of Billboards,'" *Washington University Journal of Urban and Contemporary Law* 30 (1986), p. 361; Daniel R. Mandelker, "The Free Speech Revolution in Land Use Control," *Chicago-Kent Law Review* (1984), p. 55.
69. *Metromedia, Inc., et al. v. City of San Diego et al.*, 453 U.S. (1983) 570.
70. Mandelker, "Free Speech Revolution," pp. 61–62.
71. *Members of the City Council of Los Angeles et al. v. Taxpayers for Vincent et al.*, 466 U.S. (1984) 817.
72. Sanders, "First Amendment," pp. 336–37; Daniel R. Mandelker and William R. Ewald, *Street Graphics and the Law*.
73. *Members of the City Council of Los Angeles et al. v. Taxpayers for Vincent et al.*, 466 U.S. (1984) 822.
74. Jennifer Flinchpaugh, "Legislation Matters," *Signs of the Times* 223 (Sept. 2001), pp. 76, 78.
75. "Sign Legends — Jim Claus: A Pioneer in Sign Legislation," *Signs of the Times* 216 (Mar. 1994), pp. 174, 290.
76. For example, see his series of recommendations in *Signs of the Times*: Karen E. Claus, Susan Lynn Claus, Donald Large, and Robert James Claus, "Legal Issues," 210 (Nov. 1988), pp. 52, 54, 56, 60; 210 (Dec. 1988), pp. 34, 36, 38, 40, 42, 46, 152; 211 (Feb. 1989), pp. 64, 66, 68, 70, 72; 211 (Apr. 1989), pp. 72, 74–75, 78, 80, 82, 84, 86; 211 (June 1989), pp. 40, 42, 44, 48, 50, 52, 54; 211 (Oct. 1989), pp. 30, 32, 34, 36, 40; 212 (Feb. 1990), p. 70; 212 (Dec. 1990), pp. 32, 34, 36, 38, 40, 169.
77. R. James Claus and Karen E. Claus, *Visual Environment: Sight, Sign and By-Law*, pp. 60–61.
78. Mandelker and Ewald, *Street Graphics*, p. 1; Peter H. Phillips, "Sign Controls for Historic Signs," *Signs of the Times* 211 (Apr. 1989), pp. 206, 208, 210.
79. Miller, "Great Billboard Double-Cross," p. 18; Ann Lallande, "Outdoor and Inner City," *Marketing and Media Decisions* 24 (Feb. 1989), p. 108.
80. Miller, "Great Billboard Double-Cross," pp. 14–16, 18, 20–21; Willliams, "Litter on a Stick."

EPILOGUE

1. Wilbur Zelinsky, "On the Superabundance of Signs in Our Landscape," in *Exploring the Beloved Country*, p. 277.
2. Sean O'Leary, "Digital Sign Technology: The Dynamic Signage Revolution, Part 1," *Signs of the Times* 224 (Apr. 2002), p. 62.
3. Ibid., p. 176.

Bibliography

COURT CASES

Bill Posting Sign Company v. Atlantic City, 58 Atl. (1904) 342.

Churchill and Tait v. Rafferty, 32 P.I. 39.

City of Chicago et al. v. Gunning System, 214 Ill. (1905) 628.

City of Chicago v. Gunning System, 73 N.E. (1905) 1035.

Commonwealth v. Boston Advertising Company, 74 N.E. (1905) 601.

Crawford v. Topeka, 51 Kan. (1893) 756.

Curran Bill Posting and Distributing Company v. Denver, 107 Pac. (1910) 261.

General Outdoor Advertising Company, Inc., et al. v. Department of Public Works, 289 Massachusetts (1935) 149.

The Haller Sign Works v. The Physical Culture Training School, 249 Ill. (1911) 436.

Kelbro, Inc., v. Myrick, Secretary of State, et al., 30 A. Zd (1943) 527.

Members of the City Council of Los Angeles et al. v. Taxpayers for Vincent et al., 466 U.S. (1984) 789.

Metromedia, Inc., et al. v. City of San Diego et al., 453 U.S. (1983) 490.

Passaic v. Patterson Bill Posting, Advertising and Sign Painting Co., 72 N.J.L. (1905) 285.

Perlmutter Furniture Company et al. v. Greene, Superintendent of Public Works, et al., 182 N.E. (1932) 5.

St. Louis Gunning Advertisement Co. v. City of St. Louis et al., 137 S.W. (1911) 929.

The Thomas Cusack Company v. The City of Chicago, 267 Ill. (1915) 344.

Varney and Green v. Williams (East San Jose), 100 Pac. (1909) 867.

BOOKS

Adams, Paul C. "Peripatetic Imagery and Peripatetic Sense of Place." In Paul C. Adams, Steven Hoelscher, and Karen E. Till, eds., *Textures of Place: Exploring Humanist Geographies*. Minneapolis: University of Minnesota Press, 2001, pp. 186–206.

Agnew, Hugh E. *Outdoor Advertising*. New York: McGraw-Hill, 1938.

Appleton, Jay. *The Experience of Landscape*. London: Wiley, 1975.

Appleyard, Donald, Kevin Lynch, and John R. Myer. *The View from the Road*. Cambridge, MA: MIT Press, 1964.

Armstrong, Ellis L., Michael C. Robinson, and Suellen M. Hoy, eds. *History of Public Works in the United States, 1776–1976*. Chicago: American Public Works Association, 1976.

Arthur, Paul, and Romedi Passini. *Wayfinding: People, Signs, and Architecture*. New York: McGraw-Hill, 1992.

Axia, Giovanna, Erminielda Mainardi Peron, and Marcia Rosa Baroni. "Environmental Assessment across the Life Span." In Tommy Gärling and

Gary W. Evans, eds. *Environment, Cognition, and Action: An Integrated Approach.* New York: Oxford University Press, 1991, pp. 221–44.

Barker, James F., Michael W. Fazio, and Mark Hildebrandt. *The Small Town As an Art Object.* Starkville, MS: n.p., 1975.

Barthel, Diane. *Putting on Appearances: Gender and Advertising.* Philadelphia: Temple University Press, 1988.

Baumeister, Roy F. *Identity: Cultural Change and the Struggle for Self.* New York: Oxford University Press, 1986.

Belasco, Warren J. *Americans on the Road: From Autocamp to Motel, 1910–1945.* Cambridge, MA: MIT Press, 1979.

Bennett, J. M. *Roadsides: The Front Yard of the Nation.* Boston: Stratford, 1936.

Blake, Peter. *God's Own Junkyard: The Planned Deterioration of America's Landscape.* New York: Holt, Rinehart, & Winston, 1964.

Bluestone, Daniel M. "Roadside Blight and the Reform of Commercial Architecture." In Jan Jennings, ed., *Roadside America: The Automobile in Design and Culture.* Ames: Iowa State University Press, 1990, pp. 170–84.

Borth, Christy. *Mankind on the Move: The Story of Highways.* Washington, DC: Automotive Safety Foundation, 1969.

Boston Redevelopment Authority and U.S. Department of Housing and Urban Development. *City Signs and Lights: A Policy Study.* Cambridge, MA: MIT Press, 1973.

Boulding, Kenneth E. *The Image: Knowledge in Life and Society.* Ann Arbor: University of Michigan Press, 1956.

Brouws, Jeff, photographer. Text by Bernd Polster and Phil Patton. *Highway: America's Endless Dream.* New York: Stewart, Tabori & Chang, 1997.

Buchanan, Patrick J. *The Death of the West: How Dying Populations and Immigrant Invasions Imperil Our Country and Civilization.* New York: Thomas Dunne Books/St. Martin's Press, 2002.

Canter, David. *The Psychology of Place.* London: Architectural Press, 1977.

Casey, Edward S. "Body, Self and Landscape: A Geographical Inquiry into the Place-World." In Paul C. Adams, Steven Hoelscher, and Karen E. Till, eds., *Textures of Place: Exploring Humanist Geographies.* Minneapolis: University of Minnesota Press, 2001, pp. 403–25.

Chatburn, George R. *Highways and Highway Transportation.* New York: Thomas Y. Crowell, 1923.

Claus, Karen E., and R. James Claus. *The Sign User's Guide: A Marketing Aid.* Palo Alto, CA: Institute of Signage Research, 1978.

Claus, R. James, and Karen E. Claus. *Visual Environment: Sight, Sign and By-Law.* Don Mills, Ont.: Collier-Macmillan Canada, 1971.

Claus, [R.] James, R. M. Oliphant, and Karen Claus. *Signs: Legal Rights and Aesthetic Considerations.* Cincinnati: Signs of the Times Publishing, 1972.

Cook, Guy. *The Discourse of Advertising.* London: Routledge, 1992.

Cortese, Anthony J. *Provocateur: Images of Women and Minorities in Advertising.* Savage, MD: Rowman & Littlefield, 1999.

Cross, Gary. *An All-Consuming Century: Why Commercialism Won in Modern America.* New York: Columbia University Press, 2000.

Cullen, Gordon. *The Concise Townscape*. New York: Van Nostrand Reinhold, 1971.
Dates, Jannette L. "Advertising." In Jannette L. Dates and William Barlow, eds., *Split Image: African Americans in the Mass Media*. 2nd ed. Washington, DC: Howard University Press, 1993, pp. 461–91.
Denzin, Norman K. *Symbolic Interactionism and Cultural Studies: The Politics of Interpretation*. Oxford, UK: Blackwell, 1992.
Downing, David B., and Susan Bazargan, eds. *Image and Ideology in Modern/Postmodern Discourse*. Albany: State University of New York Press, 1991.
Downs, Roger M., and David Stea. *Maps in Mind: Reflections on Cognitive Mapping*. New York: Harper & Row, 1977.
Enfield, Clifton W. "Federal Highway Beautification: Outdoor Advertising Control, Legislation and Regulation." In John W. Houck, ed., *Outdoor Advertising: History and Regulation*. Notre Dame, IN: University of Notre Dame Press, 1969, pp. 150–62.
Ericksen, E. Gordon. *The Territorial Experience: Human Ecology As Symbolic Interaction*. Austin: University of Texas Press, 1980.
Ewen, Stuart. *Captains of Consciousness: Advertising and the Social Roots of the Consumer Culture*. New York: McGraw-Hill, 1976.
Feaster, Felicia, and Bret Wood. *Forbidden Fruit: The Golden Age of the Exploitation Film*. Baltimore: Midnight Marquee Press, 1999.
Fischer, Claude S. "Ambivalent Communities: How Americans Understand Their Localities." In Alan Wolfe, ed., *America at Century's End*. Berkeley: University of California Press, 1991, pp. 79–90.
Fischer, Fred. "Ten Phases of the Animal Path: Behavior in Familiar Situations." In Aristide H. Esser, ed., *Behavior and Environment: The Use of Space by Animals and Men*. New York: Plenum Press, 1971.
Flagg, James M. *Boulevards All the Way — Maybe*. New York: Doran, 1925.
Flink, James J. *America Adopts the Automobile, 1895–1910*. Cambridge, MA: MIT Press, 1970.
———. *The Automobile Age*. Cambridge, MA: MIT Press, 1988.
Floyd, Charles F., and Peter J. Shedd. *Highway Beautification: The Environmental Movement's Greatest Failure*. Boulder, CO: Westview Press, 1979.
Forbes, T. W., Thomas E. Snyder, and Richard F. Pain. "Traffic Sign Requirements." In Highway Research Board, *Night Visibility 1963 and 1964: 10 Reports*. Highway Research Record 70. Washington, DC: Highway Research Board, 1965, pp. 48–56.
Foster and Kleiser, *Fifty Years of Outdoor Advertising, 1901–1951*. San Francisco: Foster & Kleiser, [1951].
Foster, Mark S. *Castles in the Sand: The Life and Times of Carl Graham Fisher*. Gainesville: University Press of Florida, 2000.
Fowler, Robert Booth. *The Dance with Community: The Contemporary Debate in American Political Thought*. Lawrence: University Press of Kansas, 1991.
Fox, Stephen. *The Mirror Makers: A History of American Advertising and Its Creators*. New York: William Morrow, 1984.
Francaviglia, Richard V. *Main Street Revisited: Time, Space, and Image Building in Small-Town America*. Iowa City: University of Iowa Press, 1996.

———. *The Shape of Texas: Maps As Metaphors*. College Station: Texas A&M University Press, 1995.

Francis, Mark. "Control As a Dimension of Public-Space Quality." In Irwin Altman and Ervin H. Zube, eds., *Public Places and Spaces*. New York: Plenum Press, 1989, pp. 147-71.

Fraser, James. *The American Billboard: 100 Years*. New York: Abrams, 1991.

Friedberg, Anne. *Window Shopping: Cinema and the Postmodern*. Berkeley: University of California Press, 1993.

General Outdoor Advertising Co. *Annual Report for 1957*. N.p.: General Outdoor Advertising Co., 1957.

Gerson, Kathleen. "Coping with Commitment: Dilemmas and Conflict of Family Life." In Alan Wolfe, ed., *America at Century's End*. Berkeley: University of California Press, 1991, pp. 35–57.

Gibson, James J. *The Ecological Approach to Visual Perception*. Boston: Houghton Mifflin, 1979.

Goffman, Erving. *The Presentation of Self in Everyday Life*. Garden City, NY: Doubleday Anchor, 1959.

Goldman, Robert. *Reading Ads Socially*. London: Routledge, 1992.

Golledge, Reginald G. "Cognition of Physical and Built Environment." In Tommy Gärling and Gary W. Evans, eds., *Environment, Cognition, and Action: An Integrated Approach*. New York: Oxford University Press, 1991, pp. 35–62.

Goodrum, Charles, and Helen Dalrymple. *Advertising in America: The First 200 Years*. New York: Harry N. Abrams, 1990.

Gottdiener, M. *Postmodern Semiotics: Material Culture and the Forms of Postmodern Life*. Oxford, UK: Blackwell, 1995.

Gould, Lewis L. *Lady Bird Johnson and the Environment*. Lawrence: University Press of Kansas, 1988.

Gubbels, Jac L. *American Highways and Roadsides*. Boston: Houghton Mifflin, 1938.

Hatton, Hap. *Tropical Splendor: An Architectural History of Florida*. New York: Alfred A. Knopf, 1987.

Heimann, Jim. *California Crazy and Beyond: Roadside Vernacular Architecture*. San Francisco: Chronicle Books, 2001.

Heimann, Jim, and Rip Georges. *California Crazy: Roadside Vernacular Architecture*. Introduction by David Gebhardt. San Francisco: Chronicle Books, 1980.

Hess, Alan. *Viva Las Vegas: After-Hours Architecture*. San Francisco: Chronicle Books, 1993.

Hewitt, John P. *Dilemmas of the American Self*. Philadelphia: Temple University Press, 1989.

Hokanson, Drake. *The Lincoln Highway: Main Street across America*. Iowa City: University of Iowa Press, 1988.

Howie, George W. "Traffic Signaling." In John E. Baerwald, ed., *Traffic Engineering Handbook*. Washington, DC: Institute of Traffic Engineers, 1965, pp. 350–81.

Hunter, James Davison, and John Steadman Rice. "Unlikely Alliances: The Changing Contours of American Religious Faith." In Alan Wolfe, ed., *America at Century's End*. Berkeley: University of California Press, 1991, pp. 318–31.

Ilf, Ilya, and Eugene Petrov. *Little Golden America*. London: George Routledge, 1936.

Ise, John. *Our National Park Policy: A Critical History*. Baltimore: Johns Hopkins University Press, 1961.

Jackson, J[ohn] B[rinkerhoff]. "The Future of the Vernacular." In Paul Groth and Todd W. Bressi, eds., *Understanding Ordinary Landscapes*. New Haven, CT: Yale University Press, 1997, pp. 145–54.

Jakle, John A. *City Lights: Illuminating the American Night*. Baltimore: Johns Hopkins University Press, 2001.

Jakle, John A., and Keith A. Sculle. *Fast Food: Roadside Restaurants in the Automobile Age*. Baltimore: Johns Hopkins University Press, 1999.

———. *The Gas Station in America*. Baltimore: Johns Hopkins University Press, 1994.

Jakle, John A., Keith A. Sculle, and Jefferson S. Rogers. *The Motel in America*. Baltimore: Johns Hopkins University Press, 1996.

Jhally, Sut. *The Codes of Advertising: Fetishism and the Political Economy of Meaning in the Consumer Society*. New York: St. Martin's Press, 1987.

Jones, C. Paul, and Eugene M. Wilson. "Advanced Signing for Recreational and Historic Sites." In Transportation Research Board, *Traffic Control Devices for Highways, Work Zones, and Railroad Grade Crossings*. Transportation Research Record No. 1254. Washington, DC: Transportation Research Board, 1990, pp. 8–13.

Keller, Ulrich. *The Highway As Habitat: A Roy Stryker Documentation, 1943–1955*. Santa Barbara, CA: University Art Museum, 1986.

Kelly, G. A. *The Psychology of Personal Constructs*. Vol. 1, *A Theory of Personality*. New York: W. W. Norton, 1955.

Kern-Foxworth, Marilyn. *Aunt Jemima, Uncle Ben, and Rastus: Blacks in Advertising, Yesterday, Today, and Tomorrow*. Westport, CT: Greenwood Press, 1994.

Kinnell, Galway. "The Avenue Bearing the Initial of Christ into the New World." In Mike Marqusee and Bill Harris, eds., *New York: An Anthology*. Boston: Little, Brown, 1985.

Kolb, David. "Self-Identity and Place." In *Postmodern Sophistications: Philosophy, Architecture, and Tradition*. Chicago: University of Chicago Press, 1990, pp. 146–58.

Komisar, Lucy. "The Image of Woman in Advertising." In Vivian Gornick and Barbara K. Moran, eds., *Woman in Sexist Society: Studies in Power and Powerlessness*. New York: American Library, 1972, pp. 304–17.

Lang, Jon. "Symbolic Aesthetics in Architecture: Toward a Research Agenda." In Jack L. Nasar, ed., *Environmental Aesthetics: Theory, Research, and Applications*. Cambridge, UK: Cambridge University Press, 1988, pp. 11–26.

Lay, M. G. *Ways of the World: A History of the World's Roads and of the Vehicles That Used Them*. New Brunswick, NJ: Rutgers University Press, 1992.

Leach, Neil. "Siegfried Kracauer." In Neil Leach, ed., *Rethinking Architecture: A Reader in Cultural Theory*. London: Routledge, 1997, pp. 51–52.

Le Corbusier. *When the Cathedrals Were White*. 1947. Reprint, New York: McGraw-Hill, 1964.

Lefebvre, Henri. *The Production of Space*. Translated by Donald Nicholson-Smith. Oxford, UK: Blackwell, 1991.

Leiss, William, Stephen Kline, and Sut Jhally. *Social Communication in Advertising: Persons, Products, and Images of Well-Being*. 2nd ed. Scarborough, Ont.: Nelson Canada, 1990.

Leonard, Eileen, Herman Strasser, and Kenneth Westhues, eds. *In Search of Community: Essays in Memory of Werner Stark, 1909–1985*. New York: Fordham University Press, 1993.

Lewis, Peirce F. "Common Landscapes As Historic Documents." In Steven Lubar and W. David Kingery, eds., *History from Things: Essays on Material Culture*. Washington, DC: Smithsonian Institution Press, 1993, pp. 115–39.

———. "The Geographer As Landscape Critic." In Peirce F. Lewis, David Lowenthal, and Yi-Fu Tuan, *Visual Blight in America*. Resource Paper No. 23. Washington, DC: Association of American Geographers, 1973, pp. 1–21.

Lewis, Sinclair. *Main Street*. New York: New American Library, 1961.

Liebs, Chester H. *Main Street to Miracle Mile: American Roadside Architecture*. Boston: New York Graphic Society/Little, Brown, 1985.

Lippincott, Wilmot. *Outdoor Advertising*. New York: McGraw-Hill, 1923.

Litka, Michael. "The Use of Eminent Domain and Police Power to Accomplish Aesthetic Goals." In John W. Houck, ed., *Outdoor Advertising: History and Regulation*. Notre Dame, IN: University of Notre Dame Press, 1969, pp. 89–98.

Longstreth, Richard W. *The Buildings of Main Street: A Guide to American Commercial Architecture*. Washington, DC: Preservation Press, 1987.

———. "Compositional Types in American Commercial Architecture." In Camille Wells, ed., *Perspectives in Vernacular Architecture, II*. Vernacular Architecture Forum. Columbia: University of Missouri Press, 1986, pp. 12–24.

Lowenthal, David. "The Offended Eye: Toward an Excrescential Geography." In Peirce F. Lewis, David Lowenthal, and Yi-Fu Tuan, *Visual Blight in America*. Resource Paper No. 23. Washington, DC: Association of American Geographers, 1973, pp. 29–43.

Lynch, Kevin. *The Image of the City*. Cambridge, MA: Technology Press, 1960.

Machor, James L. *Pastoral Cities: Urban Ideals and the Symbolic Landscape of America*. Madison: University of Wisconsin Press, 1987.

Mandelker, Daniel R., and William R. Ewald. *Street Graphics and the Law*. Washington, DC: Planners Press, 1988.

Marchand, Roland. *Advertising the American Dream: Making Way for Modernity, 1920–1940*. Berkeley: University of California Press, 1985.

Marcus, Leonard S. *The American Store Window*. New York: Watson-Guptill, 1978.

Margolin, Victor, Ira Brichta, and Vivian Brichta. *The Promise and the Product: 200 Years of American Advertising Posters*. New York: MacMillan, 1979.

Marling, Karal Ann. *Colossus of Roads: Myth and Symbol along the American Highway*. Minneapolis: University of Minnesota Press, 1984.

Martinson, Tom. *Signs of the City: The Role of Outdoor Advertising in Municipal Planning*. [Minneapolis]: self-published, 1995.

Marx, Leo. *The Machine in the Garden: Technology and the Pastoral Ideal in America.* 1964. Reprint, New York: Oxford University Press, 1973.

Massachusetts General Laws. [Boston]: Wright & Potter, 1932.

McBain, Howard L. *American City Progress and the Law.* New York: Columbia University Press, 1918.

McShane, Clay. *Down the Asphalt Path: The Automobile and the American City.* New York: Columbia University Press, 1994.

Meeks, Harold A. *On the Road to Yellowstone: The Yellowstone Trail and American Highways, 1900–1930.* Missoula, MT: Pictorial Histories Publishing, 2000.

Meinig, D. W. "Symbolic Landscapes: Models of American Community." In D. W. Meinig, ed., *The Interpretation of Ordinary Landscapes.* New York: Oxford University Press, 1979, pp. 164–92.

Mitchell, Arnold. *The Nine American Lifestyles.* New York: Warner, 1983.

Mitchell, Don. *Cultural Geography: A Critical Introduction.* Oxford, UK: Blackwell, 2000.

Montgomery, John A. *Eno: The Man and the Foundation: A Chronicle of Transportation.* Westport, CT: Eno Foundation for Transportation, 1988.

Moody's Investors Service. *Moody's Industrial Manual.* New York: Moody's Investors Service, 1963, 1965, 1989, 2000.

———. *Moody's Manual of Investments and Securities.* New York: Moody's Investors Service, 1927.

Nairn, Ian. *The American Landscape: A Critical View.* New York: Random House, 1965.

Newman, Richard. "Censorship in America Today." In *Proceedings of the Urban Signage Forum.* Washington, DC: Department of Housing and Urban Development, 1976, pp. 108–15.

Norris, James D. *Advertising and the Transformation of American Society, 1865–1920.* New York: Greenwood Press, 1990.

O'Barr, William M. *Culture and the Ad: Exploring Otherness in the World of Advertising.* Boulder, CO: Westview Press, 1994.

Outdoor Advertising Association of America. *Outdoor Advertising — the Modern Marketing Force.* N.p.: Outdoor Advertising Association of America, 1928.

Outdoor Advertising, Inc. *The Truth about Standardized Outdoor Advertising.* New York: Outdoor Advertising, Inc., 1931.

Patton, Phil. *Open Road: A Celebration of the American Highway.* New York: Simon & Schuster, 1986.

Pelkey, Ron E. "New Design Approach to Long-Span Overhead Sign Structures." In Highway Research Board, *Design of Sign Supports and Structures.* Highway Research Record No. 346. Washington, DC: Highway Research Board, 1971, pp. 1–10.

Pennsylvania Roadside Council. *This or That? (Scenery or Sign-ery?).* Media, PA: Pennsylvania Roadside Council, 1957.

Pipkin, John S. "Urban Geometry in Image and Discourse." In Paul L. Knox, ed., *The Design Professions and the Built Environment.* London: Croom Helm, 1988, pp. 62–98.

Pollock, Leslie S. "Relating Urban Design to the Motorist: An Empirical Viewpoint." In William J. Mitchell, ed., *Environmental Design: Research and*

Practice. Proceedings of the Environmental Design Research Conference, University of California, Los Angeles, 1972, pp. 11-1-1–11-1-8.

Porteous, J. Douglas. *Environment and Behavior: Planning and Everyday Urban Life*. Reading, MA: Addison-Wesley, 1977.

Powers, J. C., W. M. Szalaj, and R. L. Hollinger. "Breakaway Sign Testing, Phase I." In Transportation Research Board, *Roadside Safety Appurtenances*. Transportation Research Record No. 679. Washington, DC: National Academy of Sciences, 1978, pp. 38–40.

Presbrey, Frank. *The History and Development of Advertising*. Garden City, NY: Doubleday, 1929.

Priestley, J. B. *Midnight on the Desert*. New York: Harper, 1937.

Relph, E[dward]. *Place and Placelessness*. London: Pion, 1976.

Robinson, Charles Mulford. *Modern Civic Art; or, The City Made Beautiful*. 3rd ed. New York: G. P. Putnam's Sons, 1909.

Robinson, John. *Highways and Our Environment*. New York: McGraw-Hill, 1971.

Ross, Hayes E., and Lawrence J. Sweeney. "Evaluation of Bolted-Base Steel Channel Signpost." In Transportation Research Board, *Roadside Safety Appurtenances*. Transportation Research Record No. 679. Washington, DC: National Academy of Sciences, 1978, pp. 41–45.

Rowsome, Frank, Jr. *The Verse by the Side of the Road: The Story of the Burma-Shave Signs and Jingles*. Brattleboro, VT: Stephen Greene Press, 1965.

Ryden, Kent C. *Mapping the Invisible Landscape: Folklore, Writing, and the Sense of Place*. Iowa City: University of Iowa Press, 1993.

Sack, Robert David. *Human Territoriality: Its Theory and History*. Cambridge, UK: Cambridge University Press, 1986.

———. *Place, Modernity, and the Consumer's World: A Relational Framework for Geographical Analysis*. Baltimore: Johns Hopkins University Press, 1992.

Sanders, Scott Russell. "The Common Life." In William Vitek and Wes Jackson, eds., *Rooted in the Land: Essays on Community and Place*. New Haven, CT: Yale University Press, 1996, pp. 40–49.

Saroyan, William. *Short Drive, Sweet Chariot*. New York: Phaedra, 1966.

Schlereth, Thomas J. *Cultural History and Material Culture: Everyday Life, Landscapes, Museums*. Charlottesville: University Press of Virginia, 1992.

Schudson, Michael. *Advertising, the Uneasy Persuasion: Its Dubious Impact on American Society*. New York: Basic Books, 1984.

Scott, Quinta, and Susan Croce Kelly. *Route 66: The Highway and Its People*. Norman: University of Oklahoma Press, 1988.

Seely, Bruce. *Building the American Highway System: Engineers as Policy Makers*. Philadelphia: Temple University Press, 1987.

Selznick, Philip. "In Search of Community." In William Vitek and Wes Jackson, eds., *Rooted in the Land: Essays on Community and Place*. New Haven, CT: Yale University Press, 1996, pp. 195–203.

Sessions, Gordon M. *Traffic Devices: Historical Aspects Thereof*. Washington, DC: Institute of Traffic Engineers, 1971.

Shaffer, Marquerite S. *See America First: Tourism and National Identity, 1880–1940*. Washington, DC: Smithsonian Institution Press, 2001.

Shi, David E. *Facing Facts: Realism in American Thought and Culture, 1850–1920.* New York: Oxford University Press, 1995.

Sommer, Robert. *Personal Space: The Behavioral Basis of Design.* Englewood Cliffs, NJ: Prentice-Hall, 1969.

Stacey, Judith. "Background toward the Postmodern Family: Reflections on Gender, Kinship, and Class in the Silicon Valley." In Alan Wolfe, ed., *America at Century's End.* Berkeley: University of California Press, 1991, pp. 17–39.

Stephan, Walter G., and David Rosenfield. "Racial and Ethnic Stereotypes." In Arthur G. Miller, ed., *In the Eye of the Beholder: Contemporary Issues in Stereotyping.* Westport, CT: Praeger, 1982, pp. 92–129.

Stilgoe, John R. *Metropolitan Corridor: Railroads and the American Scene.* New Haven, CT: Yale University Press, 1983.

Strasser, Susan. *Satisfaction Guaranteed: The Making of the American Mass Market.* New York: Pantheon, 1989.

Sullivan, William M. "A Public Philosophy for Civic Culture." In William Vitek and Wes Jackson, eds., *Rooted in the Land: Essays on Community and Place.* New Haven, CT: Yale University Press, 1996, pp. 235–43.

Teague, Tom. *Searching for 66.* Springfield, IL: Samizdot House, 1991.

Theeuwes, Jan. "Visual Selection: Exogenous and Endogenous Control." In A. D. Gale et al., eds., *Vision in Vehicles: III.* Proceedings of the 3rd Conference on Vision in Vehicles. Amsterdam: Elsevier Science Publications, 1991, pp. 53–61.

Tocker, Philip. "Standardized Outdoor Advertising: History, Economics, and Self-Regulation." In John W. Houck, ed., *Outdoor Advertising: History and Regulation.* Notre Dame, IN: University of Notre Dame Press, 1969, pp. 11–56.

Tunnard, Christopher, and Boris Pushkarev. *Man-Made America: Chaos or Control?* New Haven, CT: Yale University Press, 1963.

U.S. Code Congressional Service. *Laws of Congress,* 78th Cong., 2nd sess. St. Paul, MN: West Publishing, 1944; Brooklyn, NY: Edward Thompson, 1945.

U.S. Department of Transportation, Federal Highway Administration. *Highway Statistics: Summary to 1985.* Washington, DC: U.S. Government Printing Office, 1985.

———. *Manual on Uniform Control Devices (MUTCD).* mutcd.fhwa.dot.gov/kno-faq.htm (accessed Feb. 2, 2002).

Venturi, Robert, Denise Scott Brown, and Steven Izenour. *Learning from Las Vegas.* Cambridge, MA: MIT Press, 1972.

Wallis, Michael. *Route 66: The Mother Road.* New York: St. Martin's Press, 1990.

Warner, Sam Bass, Jr. *Streetcar Suburbs: The Process of Growth in Boston, 1870–1900.* Cambridge, MA: Harvard University Press, 1962.

Waters, Judith, and George Ellis. "The Selling of Gender Identity." In Mary Cross, ed., *Advertising and Culture: Theoretical Perspectives.* Westport, CT: Praeger, 1996, pp. 91–103.

Watts, Margit Misangyi. *High Tea at Halekulani: Feminist Theory and American Clubwomen.* Brooklyn, NY: Carlson Publishing, 1993.

Wiley, Carroll Carson. *Principles of Highway Engineering.* New York: McGraw-Hill, 1928.

Williamson, Judith. *Decoding Advertisements: Ideology and Meaning in Advertising.* London: Marion Boyars, 1978.

Wilson, William H. *The City Beautiful Movement.* Baltimore: Johns Hopkins University Press, 1989.

Wuthnow, Robert. *Sharing the Journey: Support Groups and America's New Quest for Community.* New York: Free Press, 1994.

Zelinsky, Wilbur. *Nation into State: The Shifting Symbolic Foundations of American Nationalism.* Chapel Hill: University of North Carolina Press, 1988.

———. *Exploring the Beloved Country: Geographic Forays into American Society and Culture.* Iowa City: University of Iowa Press, 1994.

PERIODICALS

"Adverse Legislation — a Blessing in Disguise That Won Favor for Painted Display." *Signs of the Times* 44 (May 1923), p. 6.

"Advertising Investment in 1,100,000 Signs . . . More than $205,000,000 Yearly." *Signs of the Times* 84 (Dec. 1936), pp. 7–9.

"Advertising Litter Now Causing Real Barrier to Beauty." *Roadside Bulletin* 2 (May 1933), pp. 4, 8–18.

Alexander, Keith L. "Billboards Help Media Firms Weather Slowdown." *USA Today*, Dec. 12, 2000, p. 68.

"Almost a Million." *Signs of the Times* 76 (Apr. 1931).

Altheide, David L. "Identity and the Definition of the Situation in a Mass-Mediated Context." *Symbolic Interaction* 23, no. 1 (2000), pp. 1–27.

American Plywood Association. Advertisement. *Signs of the Times* 167 (May 1964), pp. 20–21.

American Society of Planning Officials. "The Elimination of Nonconforming Signs." *Report No. 209* (Apr. 1966).

Appleyard, Donald. "Why Buildings Are Known: A Predictive Tool for Architects and Planners." *Environment and Behavior* 1 (Dec. 1969), pp. 131–56.

Arroway Traffic Signal Corp. Advertisement. *Public Safety* 10 (May 1936), p. 33.

Ashe, Harold J. "How Painted Sign Display Boosts San Diego." *Poster* 19 (Jan. 1928), p. 20.

Atherton, C. A. "Electric Signs of the Future Must Make Sales: Their 'Good Will' Days Are Over." *Signs of the Times* 42 (Aug. 1922), pp. 45–53; (Sept. 1922), pp. 46–51.

"At Last — Uniform Traffic Signs." *Public Safety* 3 (Mar. 1929), pp. 3–5.

Babcock, R. Fayerweather. "Education Department." *Signs of the Times* 18 (May 1927), pp. 21–22.

Bailey, Kristin Szylvian. "Fighting 'Civic Smallpox': The Civic Club of Allegheny County's Campaign for Billboard Regulation, 1896–1917." *Western Pennsylvania Historical Magazine* 70, no. 1 (Jan. 1987), pp. 3–28.

Baker, Newman F. "Municipal Aesthetics and the Law." *Illinois Law Review* 20 (Feb. 1926), pp. 546–72.

Barker, Olivia. "'Stars and Stripes' Is Making a Lone Star Statement." *USA Today*, July 6, 2001, p. D1.

Barrett, Ross. "Fitting Outdoor Advertising into the Environment." *Signs of the Times* 197 (Oct. 1972), pp. 30–33.

Bell, Martin L., and Richard Wendel. "State Compliance to HBA Will Be Costly." *Signs of the Times* 175 (Mar. 1967), p. 107.

Bertucci, Andrew. "Electric Awning Signs Legislation and Zoning." *Signs of the Times* 223 (Jan. 2001), pp. 106–7.

Berry, Donald S. "They Shall Not Pass!" *Public Safety* 14 (Apr. 1939), pp. 20–21, 38.

Betts, James H. "Flashers Make Electrics Compelling and Lengthen Life of Lamps." *Signs of the Times* 59 (June 1925).

"Bids on Signs for Entire Store Front from Maine to California." *Signs of the Times* 50 (July 1925), pp. 82–83.

"Big One in the West." *Signs of the Times* 92 (June 1939).

Bill, Lilas S. "Posters a Factor in Good Citizenship." *Poster* 14 (Dec. 1923), pp. 5–6, 24–29.

"Billboard Industry Seeks to Vitiate Bay State Statute." *Roadside Bulletin* 2, no. 4 (May 1933), pp. 18–19.

"'Billboards Destroy Motoring Joy,' Is Successful Slogan." *Roadside Journal* 2, no. 5 (Jan. 1934), p. 3.

Blaine, Lynn B. "The Dawning of Awnings." *Signs of the Times* 184 (Aug. 1984), pp. 41–43.

Blumer, Herbert. "Sociological Implications of the Thought of George Herbert Mead." *American Journal of Sociology* 71 (1966), pp. 535–44.

"Buick's Identification Program." *Signs of the Times* 82 (Aug. 1937), p. 20.

Carr, Stephen, and Dale Schissler. "The City As a Trip: Perceptual Selection and Memory in the View from the Road." *Environment and Behavior* 1 (June 1969), pp. 7–35.

Carter, Christie. "The Beckoning Archives: Tourism Records at the Vermont State Archives." *Vermont History* 65, no. 3/4 (Summer/Fall 1997), pp. 168–70.

"The Case for Highway Adv[ertising]." *Signs of the Times* 146 (May 1957).

Chandler, Philip. "La Palina Cigars 'Made Good.'" *Poster* 16 (July 1925), pp. 7–9, 27.

"Chrysler Offers All Dealers Uniform Used Car Signs." *Signs of the Times* 151 (Jan. 1959), pp. 31, 101.

"Circulation Thirteen Cents per Thousand at the Crossroads of the World." *Signs of the Times* 50 (June 1925), p. 28.

"Cities That Advertise by Means of Slogan Signs." *Signs of the Times* 23 (Dec. 1, 1913), p. 8.

"Cities That Have Slogan Signs." *Signs of the Times* 23 (Nov. 1, 1913), p. 34.

Claus, Karen E., Susan Lynn Claus, Donald Large, and Robert James Claus. "Legal Issues," *Signs of the Times* 210 (Nov. 1988); 210 (Dec. 1988); 211 (Feb. 1989); 211 (Apr. 1989); 211 (June 1989); 211 (Oct. 1989); 212 (Feb. 1990); 212 (Dec. 1990).

Conner, Susan. "2000 Digital Service Bureau Survey." *Signs of the Times* 223 (Jan. 2001), pp. 78–80.

"Connecticut Color Route Sign System — Operation of the 'Eastern Plan.'" *Signs of the Times* 31 (Feb. 1919), p. 45.
Curry, Howard J. "New Maine Law Will Develop Outdoor Advertising to High Standard." *Signs of the Times* 51 (Sept. 1925), pp. 7, 14.
"The Dangers of Traffic: The Psychology of Accidents." *Scientific American*, supplement no. 1755 (Aug. 21, 1909), p. 127.
Danielson, Kay. "Stately Signs." *Signs of the Times* 214 (Feb. 1992), pp. 92–95.
Dascent, Bury Irwin. "Electricity As Applied to Commercial Advertising." *Journal of Electricity, Power and Gas* 16 (Feb. 1906), pp. 99–100.
Daschbach, A. C. "The Why of the Sign Board." *Signs of the Times* 27 (Oct. 1914), pp. 5, 26.
"Diminishing Traffic Law Fractures with the Aid of the Bulletin." *Signs of the Times* (Aug. 1923). p. 21.
Dorsey, Bill, and Tod Swormstedt. "History of the Sign Industry in America, 1906–1981." *Signs of the Times* 203 (May 1981), pp. 1–32.
"Do the Road Signs Leading to Your City Advertise It Properly?" *Signs of the Times* 29 (Oct. 1915), p. 26.
"Double Service Bulletins 'Guarding' Down East Roads Get Publicity for Hood Tires." *Signs of the Times* 37 (Oct. 1917), p. 26.
Dunbar, G[ary] S. "Illustrations of the American Earth: A Bibliographical Essay on the Cultural Geography of the United States." *American Studies: An International Newsletter*, supplement to *American Quarterly* 25 (1973), pp. 3–15.
Eagle Sign Corp. Advertisement. *Public Safety* 9 (May 1935), p. 27.
"Educating Drivers." *Public Safety* 5 (Sept. 1931), pp. 6–7.
"Effective Types of Danger Signals." *Concrete Highway Association* 2 (Mar. 1918), p. 93.
"Efficient, Economical." *GE Review* 42 (Mar. 1939).
Ehrenhalt, Alan. "The Empty Square." *Preservation* 52, no. 2 (Mar./Apr. 2000), pp. 43–51.
"Electronic Signs Aid Traffic Control." *Signs of the Times* 196 (Jan. 1974), pp. 46–48.
Fairley, E. M. "Poster People." *Poster* 11 (Feb. 1920).
Farnsworth, C. S. Untitled verse. *Roadside Bulletin* 2, no. 3 (c. 1933), p. 16.
"Finalized Federal Standards Are Announced." *Signs of the Times* 150 (Dec. 1958), pp. 52–54.
Flinchpaugh, Jennifer. "Legislation Matters." *Signs of the Times* 223 (Sept. 2001), pp. 76, 78.
Flinchpaugh, Jennifer, and Linda Kitchen. "Flex in the City (and Everywhere Else)." *Signs of the Times* 222 (Sept. 2000), pp. 106–13.
Foley, A. V. "Experience of the Tide Water Oil Company in the Use of Outdoor Advertising." *Signs of the Times* 39 (Nov. 1921), pp. 9–10, 74.
"Forward America." *Signs of the Times* 64 (Jan. 1930), p. 94.
Foster and Kleiser. Adverisement. *Poster* 14 (Apr. 1923), p. 3.
Francis, E. L. "Structural Standardization in Outdoor Advertising." *Advertising Outdoors* 19 (July 1928), pp. 7–10.
Fulge, Francis M. "Architectural Sign Lighting Developments." *Signs of the Times* 71 (May 1932).

Fulton, Kerwin H. "From Fencepost Daubs to De Luxe Showings Is Advance Made in Poster Advertising." *Signs of the Times* 41 (July 1922), pp. 6–8, 17.

Gamst, Frederick C. "Rethinking Leach's Structural Analysis of Color and Instructional Categories in Traffic Control Signals." *American Ethnologist* 2, no. 2 (May 1975), pp. 271–95.

Gardner, George K. "The Massachusetts Billboard Decision." *Harvard Law Review* 69, no. 6 (Apr. 1936), pp. 869–902.

Glassie, Henry. "Structure and Function, Folklore and Artifact." *Semiotica* 7 (1973), pp. 313–51.

"GOA, F & K Contend for No. l Rank in Outdoor Field; Naegle Moves Up." *Advertising Age* 34 (July 1, 1963), p. 79.

Goodrich, Chauncey Shafter. "Billboard Regulation and the Aesthetic Viewpoint with Reference to California Highways." *California Law Review* 17 (1928–29), pp. 120–34.

Gordon, D. Logan, and N. Jane Zbrodoff. "Selection for Cognition: Cognitive Constraints on Visual Spatial Attention." *Visual Cognition* 6 (1999), pp. 55–81.

Green, Justin. "Sign Game." *Signs of the Times* 223 (Apr. 2001), p. 9.

Grieves, Thomas O. "What Posters Did for a Church." *Poster* 15 (Jan. 1924), pp. 26–27.

Halsted, J. N. "Modern Sign Making with Lacquer." *Signs of the Times* 60 (July 1930), pp. 24–26.

Harrington, Burton. "What Is Poster Advertising?" *Poster* 14 (Oct. 1923), pp. 3–4, 14–15, 30.

Harris, Paul B., and Barbara B. Brown. "The Home and Identity Display: Interpreting Resident Territoriality from Home Exteriors." *Environmental Psychology* 16 (1996), pp. 187–203.

Henderson, Cliff. "Route Signs . . . Yesterday and Today." *Signs of the Times* 148 (Jan. 1958), pp. 74, 100.

Henderson, Roberta. "The Enduring Curse of Billboard Blight." *Louisville Courier Journal*, Nov. 2, 1997, p. D-3.

Herbert, Olga. "Where the Rubber Meets the Road." *Lincoln Highway Journal* 2, no. 1 (Summer 1997), p. 2.

Hewes, L. I. "Control Roadside Advertising and Aesthetics." *Roads and Streets* 69 (Oct. 1929), pp. 346–48.

Highway Lighthouses. Advertisement. *Public Safety* 43 (June 1931), p. 43.

"Highway Safety Signs." *Signs of the Times* 85 (Jan. 1937), p. 96.

Hirst, A. R. "How Wisconsin Marks Its Highways — 5,000 Mile System Now Signed." *Signs of the Times* 31 (Feb. 1919), p. 44.

Holiday, S. N. "Through the Years with Electrical Advertising on the Great White Way." *Signs of the Times* 68 (May 1931), pp. 30–31, 56–57.

"How California Marks Its Highways — 60,000 Direction Signs Placed." *Signs of the Times* 31 (Mar. 1919), p. 49.

Igert, M. "The Psychological Laws of Posters, Part I." *Poster* 18 (Mar. 1927), pp. 18–20, 28.

"Illumination Is the Symbol of Cheer and the Sign of Patriotism." *Signs of the Times*, Apr. 1918, pp. 34–35.

"Iowa OAA Goes All Out to Win Friends, Serve Public." *Signs of the Times* 150 (Dec. 1958), pp. 40–41, 108.

Jacks, Leroy. "Electrical Advertising Opportunities in Small Cities." *Signs of the Times* 15 (Nov. 1, 1913), p. 10.

Jakle, John A. "Pioneer Roads: America's Early Twentieth-Century Named Highways." *Material Culture* 32, no. 2 (2000), pp. 1–22.

Jersig, W. H. "San Antonio's Miniature Bulletin Boards Are Dedicated to the Weary Traveler." *Signs of the Times* 15 (Dec. 1923), p. 12.

Jones, Charissa. "Leaving Their Mark Behind." *USA Today*, Nov. 22, 2000, p. 19A.

Kelley, E. Thomas. "Changing Habits of Americans Rapidly Increasing Outdoor Circulation." *Signs of the Times* 47 (June 1927), p. 11.

———. "Signs of the Times Celebrates Its Twentieth Year of Service." *Signs of the Times* 52 (May 1926), pp. 5–6.

"Kennedy Takes on the Billboard Barons." *Christian Century*, Apr. 12, 1961, p. 147.

King, William J. "Meeting Destruction with Construction — Opposition without Opposition." *Signs of the Times* 44 (May 1923), pp. 8–9, 20–21.

Kline, Theresa J. B., Laura M. Ghali, and Donald W. Kline. "Visibility Distance of Highway Signs among Young, Middle-Aged, and Older Observers: Icons Are Better than Text." *Human Factors* 32 (Oct. 1990), pp. 609–19.

Klotter, James C., and John W. Muir. "Boss Ben Johnson, the Highway Commission, and Kentucky Politics, 1927–1937." *Register of the Kentucky Historical Society* 84, no. 1 (Winter 1986), pp. 18–50.

Kriemm, George. "Chicago Made Hideous with Signboards." *Chicago Sunday Times-Herald*, May 6, 1900, part 4, page 3.

Lallande, Ann. "Outdoor and Inner City." *Marketing and Media Decisions* 24 (Feb. 1989), p. 108.

"Largest Service Station Sign Erected by Halfer and Ragg." *Signs of the Times* 52 (Mar. 1926), p. 54.

Lemperly, C. M. "A Bulletin Campaign of Giant Trade-Mark Cutouts." *Signs of the Times* 73 (June 1935).

Lewis, Mary Beth. "Ten Best Road Signs." *Car and Driver* 35, no. 7 (Jan. 1990), p. 118.

Lewis, Peirce F. "Learning from Looking: Geographic and Other Writing about the American Landscape." *American Quarterly* 35 (1983), pp. 242–61.

Ley, David, and Roman Cybriwsky. "Urban Graffiti As Territorial Markers." *Annals, Association of American Geographers* 64 (Dec. 1974), pp. 491–505.

Lloyd, M. G. "Uniform Traffic Signs, Signals, and Markings." *Annals of the American Academy of Political and Social Science* 133 (Sept. 1927), pp. 121–27.

Loewe, Arthur F. "Selling Safety on the Highways." *Signs of the Times* 85 (Mar. 1937), p. 46.

Logan, Gordon D., and N. Jane Zbrodoff. "Selection for Cognition: Cognitive Constraints on Visual Spatial Attention." *Visual Cognition* 6 (1999).

Lucht, Harry. "Main Street, USA." *Roadside Bulletin* 2 (May 1933), pp. 22–23.

"Lurid Billboards Contribute to Annual Loss of Lives by Accidents on Highways." *Roadside Bulletin* 2 (May 1933), p. 1.

Maguire, Megg, and Frank Vespe. "'I Think That I Shall Never See a Billboard Lovely As a Tree.'" *Forum Journal* 14, no. 4 (Summer 2000), pp. 57–64.
Mandelker, Daniel R. "The Free Speech Revolution in Land Use Control." *Chicago-Kent Law Review* (1984), pp. 51–62.
Mattson, Richard. "Store Front Remodeling on Main Street." *Journal of Cultural Geography* 3 (Spring/Summer 1983), pp. 41–55.
McCarthy, A. L. "Eureka Launches National Campaign." *Poster* 15 (Apr. 1924).
McKenzie, Wilma. "Merchandising to Women." *Advertising Outdoors* 2 (Aug. 1931), pp. 7–10, 35–36.
McPherson, Orville. "San Diego Welcomes Tourists with Bulletins along the Highways." *Signs of the Times* 58 (Apr. 1928), p. 12.
Meier, Barry. "Lost Horizons: The Billboard Prepares to Give Up Smoking." *New York Times*, Apr. 19, 1999, pp. A1, A20.
Menefee, H. C. "W. H. Donaldson, Founder and Former Owner of Signs of the Times Dead." *Signs of the Times* 51 (Sept. 1925), pp. 8, 15.
Meyer, Robert L. "The Council between the Wars." *Safety and Health* 135, no. 5 (May 1987), pp. 22–23.
———. "A Look at the First Congress." *Safety and Health* 135, no. 2 (Feb. 1987), pp. 22–23.
Miller, James Nathan. "The Great Billboard Double-Cross." *Reader's Digest* 126 (July 1985), pp. 83–90.
"Millions of Americans Depend on Highway Advertising." *Signs of the Times* 145 (Feb. 1957), pp. 84, 86–87, 138.
"A Model Bulletin Sign." *Signs of the Times* 9 (Nov. 1909), p. 24.
Mueller, E[dward] A. "Aspects of the History of Traffic Signals." *IEE Transactions on Vehicular Technology* 19, no. 1 (Feb. 1970), pp. 6–17.
———. "The Transportation Profession in the Bicentennial Year, Part II." *Traffic Engineering* 46 (Sept. 1976), pp. 29–34.
"National Advertising at the Point of Sale." *Signs of the Times* 79 (Mar. 1935).
"National Legislation Proposed for Road Signs — Congress Asked to Pass Protection Law." *Signs of the Times* 32 (Nov. 1919), p. 65.
Neal, Harry E. "Proper Marking of Highways Essential to Attain Maximum Services." *Signs of the Times* 51 (Dec. 1925), p. 6.
"New Road Signs to Make Ohio Roads Best Marked in Country." *Signs of the Times* 44 (July 1923), pp. 60–61.
"New Type of Reflecting Highway Signs." *Signs of the Times* 132 (Oct. 1952), pp. 39, 139.
"New York's New Street Signs." *Electrical Age* 27 (July 1902), p. 272.
"'97 Lincoln Highway Road Rally." *Lincoln Highway Journal* 2, no. 2 (Fall 1997), p. 4.
Nokes, Tom. "Helpful Suggestions for Cleaning Up Outdoor Displays." *Signs of the Times* 49 (Jan. 1925), pp. 5–6, 21, 25.
"North Carolina's Road Signs — 'Clock System' Used." *Signs of the Times* 31 (Apr. 1919), p. 45.
"O.A.A. Votes to Back Manufacturers in Campaign of Industrial Leaders to Promote Faith in 'the American Way.'" *Signs of the Times* 84 (Nov. 1936), p. 9.

O'Leary, Sean. "Digital Sign Technology: The Dynamic Signage Revolution, Part 1." *Signs of the Times* 224 (Apr. 2002).

"Opportunity, Progress and Prosperity Symbolized by the Slogan Sign — a City's Best Ad." *Signs of the Times* 35 (Feb. 1917), pp. 24–25.

"Outdoor Advertising for Automobiles." *Signs of the Times* 33 (Sept. 1, 1913), p. 3.

"The Outdoors." *Time*, Mar. 29, 1968, p. 27.

Parsons, James J. "Hillside Letters in the Western Landscape." *Landscape* 30, no. 1 (1988), pp. 15–23.

Parsons, Russ, Louis G. Tassinary, Roger S. Ullrich, Michelle R. Hebl, and Michelle Grossman-Alexander. "The View from the Road: Implications for Stress Recovery and Immunization." *Environmental Psychology* 18 (1998), pp. 113–40.

Perry, W. T. "A New Way to Control Traffic on Congested City Streets." *American City* 22 (1920), p. 476.

Phillips, Peter H. "Sign Controls for Historic Signs." *Signs of the Times* 211 (Apr. 1989), pp. 206, 208, 210.

"The Preservation of Scenic Highways." *Poster* (Dec. 1929), pp. 20–23.

Prince, J. A. "Speed in Conveying Message Becomes More Vital Each Year." *Signs of the Times* 147 (Oct. 1957), pp. 42–43.

"Proposed Laws Threaten Rural Property Owners." *Signs of the Times* 131 (Aug. 1952), pp. 40–41.

"Proposed Statutes Reflecting Public Demand for Action." *Roadside Bulletin* 2, no. 1 (1932), pp. 1, 8.

Proshansky, Harold M., Abbe K. Fabian, and Robert Kaminoff. "Place-Identity: Physical World Socialization of the Self." *Journal of Environmental Psychology* 3 (1983), pp. 57–83.

"Quantity Signs: Progressive Years." *Signs of the Times* 143 (May 1956), pp. 158–61.

Rafferty, Edwin C. "Orderly City, Orderly Lives: The City Beautiful Movement in St. Louis." *Gateway Heritage* 11, no. 4 (Spring 1991), pp. 40–62.

Railway Station Advertising. Advertisement. *Signs of the Times* (Feb. 1911), p. 19.

"Record-Breaking Sign Order for Schlitz Beer." *Signs of the Times* 76 (Mar. 1934), pp. 19, 104.

Reese, Charley. "Flag Should be Treated As Something Extra Special." *Danville (IL) Commercial News*, Nov. 28, 2001, p. 6A.

"Reflecting Warning Signs." *American City* 45 (Mar. 1931), p. 173.

Riehle, Theodore M. "No Billboards in Vermont." *American City* 83 (Sept. 1968), p. 166.

"Road Bulletins Increase Sales Over Seven Times for Hood Tires in New England." *Signs of the Times* 28 (Apr. 1918), p. 44.

"Road Development Does Outstanding Beautification Work." *Roadside Bulletin* 2 (Jan. 1934), pp. 1, 8.

Roberts, Kenneth L. "Travels in Billboardia." *Saturday Evening Post*, Oct. 13, 1928, pp. 24–25, 186, 189–90.

Rodda, Clinton. "The Accomplishment of Aesthetic Purposes under the Police Power." *Southern California Law Review* 27 (1954), pp. 149–79.
Rucker, Evelyn C. "Women and the Poster." *Poster* 17 (May 1926), p. 13.
Salmon, Thomas W., II. "Messages along the Roadside Landscape: A Vanishing Art Remembered with Nostalgia." *APT Bulletin* 30 (1999), pp. 62–63.
Sanders, Brad. "The First Amendment 'Law of Billboards.'" *Washington University Journal of Urban and Contemporary Law* 30 (1986), pp. 333–63.
"Senate Committee Eliminates Ban of Outdoor Signs from Road Bill." *Signs of the Times* 140 (June 1955), p. 27.
"Sign Legends—Jim Claus: A Pioneer in Sign Legislation." *Signs of the Times* 216 (Mar. 1994), pp. 174, 290.
"Signman's Ingenuity Creates a Spectacular." *Signs of the Times* 133 (Feb. 1933), pp. 23–24.
"Signs Show the Way." *Lincoln Highway Journal* 2, no. 1 (Summer 1997), p. 3.
Simross, Lynn. "A Hollywood Cliff-Hanger." *Los Angeles Times*, Jan. 12, 1977, pp. 1, 7.
———. "Man behind the Hollywood Sign." *Los Angeles Times*, Jan. 18, 1977, part 4, pp. 1, 6.
Sitwell, O. F. G., S. E. Olenka, and S.E. Bilash. "Analyzing the Cultural Landscape as a Means of Probing the Non-Material Dimension of Reality." *Canadian Geographer* 30 (1986), pp. 132–45.
"Slogans and Rhymes on Signs Help Keep Motorists Alert." *Signs of the Times* 145 (Feb. 1957), p. 41.
Smith, Patrick Knox. "Laws of Layout in Display Art." *Signs of the Times* 87 (Oct. 1937).
Snow, J. Todd. "The New Road in the United States." *Landscape* 17 (Autumn 1967), pp. 13–16.
[Souder, David M.] "A Glorious Half Century of Outdoor Advertising." *Signs of the Times* 143 (May 1956), pp. 112–53.
Spencer, Nelson S. "Street Signs and Fixtures." *Municipal Affairs* 5 (Sept. 1901), pp. 426–27.
Standard and Poor's Corp. "Gannett Co., Inc." *Standard Corporation Descriptions* 59, no. 22 (Nov. 25, 1998), sec. 2.
Stark, Charles W. "New York's Early Traffic Signal Towers." *Traffic Engineering* 39 (Nov. 1968), pp. 48–49.
Stevenson, Charles. "The Great Billboard Scandal of 1960." *Reader's Digest* 76 (Mar. 1960), pp. 146–56.
Stevenson, Richard W. "Challenging the Billboard Industry." *New York Times*, Aug. 30, 1988, pp. 1, 38.
"Studebaker Uses Signs Profusely—and Why" *Signs of the Times* 28 (Mar. 1915), p. 3.
"The Structures of Organized Outdoor Advertising." *Advertising Outdoors* 21 (Dec. 1931), pp. 26–30, 42.
"Super Markets Step Up Their Use of Advertising Displays." *Signs of the Times* 137 (July 1954), pp. 23–26.

"Systematic Marking of Wayne County Highways." *Concrete Highway Magazine* 2 (Oct. 1918), p. 239.

"$10,000 Donation Saves Landmark Hollywood Sign." *Los Angeles Times*, Mar. 14, 1973, part 2, p. 1.

"3,500 Signs Placed in the West by Auto Club of Southern Cal." *Signs of the Times* 48 (Nov. 1924), p. 66.

Tilford, Kristin. "Signs of the Times." *Westways* 91, no. 4 (July/Aug. 1999), pp. 26–30.

Townsend, A. C. "If You Are Feeling Effects of the Sign Rental Bogey . . . Why Lease?" *Signs of the Times* 74 (July 1933), p. 17.

"Traffic Experts Analyze 'Through Streets.'" *Public Safety* 4 (Jan. 1930), pp. 7–8.

"A Traffic Light That Stays in Place." *American City* 24 (Apr. 1921), p. 443.

"Traffic Signs Offer Opportunities to Sign Men — How They Aid in Making Service Complete." *Signs of the Times* 37 (Oct. 1917), p. 38.

Twigger-Ross, Clare L., and David L. Uzzell. "Place and Identity Processes." *Environmental Psychology* 16 (1996), pp. 205–20.

Tymoski, John. "Park Service Goes Plastic." *Signs of the Times* 217 (Nov. 1995), pp. 132–35.

Van Duzer, W. A. "Don't Walk! Walk!" *Public Safety* 20 (July 1941), pp. 20–21.

"Victory for 'Sky Signs' in N.Y." *Signs of the Times* 9 (May 1909) p. 7.

Warner, John DeWitt. "Advertising Run Mad." *Municipal Affairs* 4 (June 1900), pp. 267–93.

Weightman, Barbara A. "Sign Geography." *Journal of Cultural Geography* 9, no. 1 (Fall/Winter 1988), pp. 53–70.

Wescott, Harris. "If Standardized Outdoor Advertising Did Not Exist." *Poster* 19 (Mar. 1928), pp. 8–11.

"What I Learned from a Poster Board." *Poster* 15 (May 1924), pp. 11–12.

Wikle, Thomas A. "Those Benevolent Boosters: Spatial Patterns of Kiwanis Membership." *Journal of Cultural Geography* 17, no. 1 (1996), pp. 1–19.

Wiley, C. C. "Road Signs for Night Driving." *Roads and Streets* 76 (Feb. 1933), pp. 75–77.

Wilks, Hugh Russell. "Detroit Insurance Company Becomes Famous through Humorous Electrical Advertising." *Signs of the Times* 48 (Dec. 1924), pp. 31–32.

Williams, Arthur. "Broadway — A Fascinating Electric Sign Gallery — America's Brightest Street." *Signs of the Times* 35 (Mar. 1917), pp. 5, 23.

Williams, George. "Electricity for National Advertising." *Signs of the Times* 9 (Aug. 1909), p. 3.

Williams, Sidney J. "Finding the Causes of Accidents." *Annals of the American Academy of Political and Social Science* 133 (Sept. 1927), pp. 156–60.

Williams, Ted. "Litter on a Stick." *Audubon* 93 (Aug./Sept. 1991), pp. 14–16, 18, 20–21.

Wilson, Ruth I. "Billboards and the Right to Be Seen from the Highway." *Georgetown Law Journal* 30 (1942), pp. 723–50.

Wilson, William H. "The Billboard: Bane of the City Beautiful." *Journal of Urban History* 13 (Aug. 1987), pp. 395–405.
"The World's Largest Spectacular." *Signs of the Times* 32 (Mar. 1917), p. 9.
Wright, James C. "Can the United States Have Both Beauty and Business?" *Signs of the Times* 177 (Dec. 1967), pp. 40–41.
Young, Frank W. "Review Essay: Putnam's Challenge to Community Sociology." *Rural Sociology* 66, no. 3 (2001), pp. 468–74.
Young, G. H. S. "Electrical Advertising in Boston." *Signs of the Times* 19 (Nov. 1912), p. 24.
"Zoning Plan Presented at Roadside Beauty Conference." *Advertising Outdoors* 2 (Apr. 1931), pp. 29, 31.
Zuhradnik, Rich. "Outdoor Obsession." *Marketing and Media Decisions* 21, no. 3 (Mar. 1986), p. 4.

Index

advertising: and advertisers, xxix, 7, 19–24, 33–37, 40–47, 49, 62, 96–107, 133, 144, 151, 153–55, 161; agencies, 7, 12; with signs, xix, xxi, xxiii–xxix, 6–9, 21–24, 33–51, 95–99, 105–07, 113, 122–24, 130–37, 146, 167, 169
Advertising Painter's League of America, 12
African Americans, 104–05, 165
American Association of Highway Officials, 56, 60–62, 66, 70
American Civic Association, 136
American Engineering Standards Committee, 57, 65
American Motor Hotel Association, 159
American Nature Association, 156
American Planning Association, 165
Appleton, Jay, 122–23
Appleyard, Donald, 120, 123
architecture, xvii, xxii–xxiv, xxix, 3, 20, 28, 42–45, 120, 122–23
Arlington, VA, 78
Associated Billposters and Distributors of the United States and Canada, 151
Associated Billposters' Association, 10, 12
Association of County Highway Engineers, 65
Association of National Advertisers, 33
Association of Window Trimmers of America, 27
Atlanta, GA, 137
Atlantic City, NJ, 85, 148
Automobile Club of Southern California, 66–67, 81
automobiles and automobility, xvii–xviii, 3, 5, 16, 19, 31–32, 39, 45, 49–51, 56, 60, 67, 80–84, 108–09, 123–32, 167

Bardstown, KY, 81
Barney Link Fund, 33
Bemidji, MN, 44
Bennett, J. M., 130
billboards, xxiii, xxx, 6–12, 37–43, 45, 47–50, 57, 81, 98–100, 106, 124, 130–33, 136–65, 168; in Georgia, 137; jumbos, 159; lizzies, 41–42; in Maine, 43, 153, 160; in Massachusetts, 153–54, 156; pilasters, 41–42; in Vermont, 156, 158, 160
Blake, Peter, 157
Bonus Act of 1958, 157–58
Booneville, KY, 122
boosterism, 80–82, 91
Boston, MA, 84, 123, 148, 154
Boulding, Kenneth, 119
Bradbury and Houghteling Company, 12
Bradford, William, ix–x
Brennan, William J., Jr., 162
Brewster, William, x
Brooklyn, NY, 13, 151
Brown, Denise Scott, 44
Buhl, MN, 89
Bureau of Street Traffic Research, 33
Burma Shave, 40–41, 51

Cannon Falls, MN, 44
Cape Cod, MA, ix–x
Carter, Garnet, 37
Casper, WY, 20–21
cemetery markers, 110–11, 113

chain stores, 21–23
Challis, ID, xxv
Champion Spark Plugs, 42
Chandler, Harry, 86
Charleston, SC, xiii
Chicago, IL, 13, 23, 148–49
Chisholm, MN, 87
Churchill and Tait v. Rafferty, 147–48
Cincinnati, OH, xxvi, 13, 106
City Beautiful movement, 135, 145–46, 148–50
Civil War, 4, 76
Claus, Karen, xxx, 47
Claus, R. James, xxx, 47, 164
Cleveland, OH, 35
Clinton, William Jefferson, xiv
Coalition for Scenic Beauty, 161
Coca-Cola, 23, 33, 105
cognitive maps, 119–23
commercial strips, xix, 39–49
Commonwealth v. Boston Advertising Company, 153
community, xix, xxxii, 17, 31, 73–91, 169
Concord, MA, 154
Conference on Roadside Business and Rural Beauty, 155
Congress Cigar Company, 99
consumerism, 6, 19, 25–26, 31, 33, 90, 95–99, 105–06
corporate logos, xxxi, 13–15, 21–24, 34–35, 98, 132, 134
Covington, KY, 6
Cross, Gary, 19, 33
Cusack, Thomas, 13, 59, 149–50, 152

Dana, IN, 78
Dasent, Bury, 5
dealer identification signs, 22–24
Decatur, GA, 137
Denver, CO, 148
Detroit, MI, 12, 39, 43, 64, 85
Dixie Highway, 81
Donaldson, William H., 12
downtown business districts, xvii–xviii, xxi–xxiii, 3–18, 21, 26, 85, 131, 136
Durham, NC, 85

Eastman Kodak Company, 7
economic depression of the 1930s, 24, 27–29, 32–34, 48, 73, 131
Edison Electric Company, 13
electric signs, 4–5, 8, 13–16, 18–19, 23, 45–46, 130; backlit canopies, 30–31; backlit panel, 16, 48–49, 168; spectaculars, 13–16, 45–46, 130, 147, 149
Eller Media, xxx
Eno, William Phelps, 56
ethnicity and race, xii, 95, 104–05
Eureka Vacuum Cleaner Company, 10–11
Everett, Aughenbaugh and Company, 25

Fairley, E. M., 98–99
family life, 76, 89–91, 108
Farm Security Administration, 131–32
fascia signs, 4, 20, 27, 29–30
fast food restaurants, xxix, 41, 44, 49, 131
Federal Highway Acts: of 1916, 58; of 1921, 59; of 1944, 62; of 1956, 157
Federal Highway Administration, 62
Federal Highway Beautification Act of 1965, 159–64
Federal Housing Act of 1935, 27
Federal Writer's Program, xiii
Fischer, Claude, 80
Fischer, Fred, 126
Fisher, Carl, 81
Flagg, James, 126
flags and banners, 8, 76–79, 91
Flashtric Sign Works, 23
Flink, James, 56
Floyd, Charles, 160
Ford Motor Company, 19

Foster and Kleiser Company, 13, 40–41, 47, 79, 156
Fowler, Robert Booth, 74–75
Francaviglia, Richard, 45, 86
Frankfort, IN, xxxii
Franklin, PA, 70
Friedberg, Anne, 25
Fulton, Kerwin H., 34, 37, 151

Galveston, TX, 85
Gamble-Skogmo, Inc., 47
Gannett Outdoor Advertising Company, 47
gasoline stations, xxix, xxxi–xxxii, 34–35, 41, 49, 131
gateway signs, 86–88
gender, 99–105, 166
General Electric Company, 13
General Motors Corporation, 19, 24
General Outdoor Advertising Company, 13, 34, 47, 154
General Outdoor Advertising Company, Inc., et al. v. Department of Public Works, 154
geography and geographers, xvii, xxii–xxiv, 55, 87, 100, 106–07, 120, 142, 163, 168
Gibson, James, 121
Goldman, Robert, 97
Golledge, Reginald, 120
graffiti, 108, 110
Great Atlantic and Pacific Tea Company, 21
Greenpeace, 165
Gunning, R. J., 13

H. J. Heinz Company, 146
Halter and Rugg Sign Company, 35
Hess, Alan, 45
highway beautification, 46–47, 159–64
Highway Displays, Inc., 35
highway safety, 56–65, 67, 69, 71, 140–41, 152
highways, 39–41, 55–63, 65–72, 80–84, 124–26, 139–40, 142, 153–57, 167
Holiday, S. N., 14
Hood Tire Company, 40, 49
Hoover, Herbert, 57, 60, 66
Hope, AR, xiv
Houston, TX, 64

ideology, xxix, 97
Ilf, Ilya, 131
Independent Grocers Alliance, 21
Indianapolis, IN, 113–14
individualism, 73–76, 80, 95, 105
Infinity Outdoor, xxx
International Bill Posters Association of North America, 12
International Road Congress, 66
Interstate Highway Sign Company, 70
Izaak Walton League, 165
Izenour, Steven, 44

J. Walter Thompson, Inc., 7
Jackson, J. B., 84
Jackson, MI, 37
Jersey City, NJ, 15
Johnson, Ben, 81
Johnson, Lady Bird, 159
Johnson, Lyndon B., 159
Joint Board on Interstate Highways, 60
Joseph Schlitz Brewing Company, 23

Kansas City, MO, 13, 45, 81
Kelley, R. Leslie, 86
Kennedy, John F., 158
Kerr, Robert S., 158
Kissam and Allen Company, 12
Komisar, Lucy, 103–04
Kracauer, Siegfried, 73
Kroger Company, 21

La Palina Cigars, 99–100
Lamar Advertising Company, 47

landscape: defined, ix–x, xvii–xix, xxi, 118–19, 134–35, 142; at night, xxiii, 5, 14–16, 29, 42, 56–57, 131; reading the, ix–xi, xvii–xviii, xxi–xxiii, 98, 106–07, 117–20, 132, 168–69; signatures of, x–xi, xxiii, 66, 75, 120, 126, 132–33, 144, 164, 168; visualization of, xxi–xxii, 56–57, 117–44, 158, 163–64
Las Vegas, NV, 45, 168
Lawton, Elizabeth Boyd, 153–54
Le Corbusier, Charles, 63
Leiss, William, 8
Lewis, Sinclair, 18
Liberal, KS, 45
liberalism, 75, 79–80, 89
lifestyle, 8, 95, 105–07
Lincoln Highway, 59, 61, 69, 81, 83
Link, Barney, 151
Little Rock, AR, 70
Loewy, Raymond, 48
London, UK, 63
Los Angeles, CA, 67, 81, 85–86
Los Angeles Cultural Heritage Board, 86
Louisville Automobile Club, 81
Louisville, KY, 13, 15, 81
Lowenthal, David, 135
Lucas, Scott, 157
Lucht, Harry, 136
Lynch, Kevin, 119–20, 123

MacFarland, J. Horace, 146, 164
Mandler, J. M., 119
Marchand, Roland, 98, 106
Marietta, OH, 82
material culture, xxi, xxix, xxxii, 49–50
McClintock, Miller, 33
McKenzie, Wilma, 102
Mead, George, xxvii
Meinig, Donald, 73
Members of the City Council of Los Angeles et al. v. Taxpayers for Vincent et al, 163
Metromedia, Inc., v. San Diego, 162
Meyer, John, 123
Miami Beach, FL, 81
Millersville, IL, 112
Milwaukee, WI, 64
Minneapolis, MN, 13, 18
Mississippi Valley Association of State Highway Departments, 60
Mitchell, Arnold, 105
Mitchell, Don, 105
modernism, 27–31, 42, 44, 50, 59, 83
Moses, Robert, 157
motels, xxix, 44–45, 49
Mulberry Castle, SC, xiii–xv

Nairn, Ian, 121
National Association of Display Industries, 27
National Association of Manufacturers, 79
National Association of Specialty Advertising Salesmen, 27
National Bureau of Standards, 59
National Conference on Street and Highway Safety, 57, 62
National Council for Industrial Safety, 56
National Council for the Protection of Roadside Beauty, 136, 153, 156, 161
National Electric Sign Association, 47
National Outdoor Advertising Bureau, 12
National Safety Council, 57
National Trust for Historic Preservation, 85
neighborhoods, 73, 84, 91, 95, 106, 110, 150, 166
Neuberger, Richard L., 156
New Netherlands, xi
New York City, NY, 3, 7, 12, 14, 34, 63–65, 78, 108, 131, 146–47, 151; Times Square, 14–15, 131
Newman, Richard, 141

O. J. Gude Company, 13, 146

Ogden, UT, 153
Omaha, NE, 13, 109
Outdoor Advertising Association of America, 13, 33, 35, 41, 45, 49, 78, 141, 151, 159
Outdoor Advertising, Inc., 34
outdoor advertising industry, xxx, 6–7, 10–13, 32–49, 137–40, 144, 149–55, 159, 162, 164–66
Outdoor Circle, 153
Outdoor Systems, Inc., 47

painted wall signs, 8–9, 12–13, 36–37
Panama-California Exposition of 1915–16, 16, 81
Parson, James, 87
Passaic, NJ, 148
Patrick Media Group, 47
Pennsylvania Roadside Council, 137
Perlmutter Furniture Company et al. v. Greene, Superintendent of Public Works, et al., 155
personal identity, xix, xxvii–xxix, xxxii, 17, 31, 74, 95, 97–100, 107–13, 169
Petrov, Eugene, 131
Philadelphia, PA, 3, 99, 109
Phoenix, AZ, 81
pictorialization, xvii, xxvii, 5, 9, 36, 97–98, 106, 119, 133
Piggly Wiggly supermarkets, 21
Pipkin, John, 119
Pittsburgh Reflector and Illuminating Company, 29
place: constructs of, xxvii, 98–101, 117–22; cues to, ix–xi, xxvi–xxvii, xxi, xxxi, 8, 18, 121–26, 143, 163; defined, xvii–xix, xxi–xxvi; image of, xxiv, xxvi–xxviii, 8, 26; meaning of, ix–xiii, xxi, xxiv, xxvi–xxviii, xxxi–xxxii, 4, 18, 95, 98–101, 117–20, 167, 169; sense of, xxvii, xxxii, 88–89, 131
place-product-packaging, 22–23
point-of-purchase signs, 23–24, 31, 34
Porter, IN, 36
Portland, OR, 13
Poster Advertising Association, 85
Poster Advertising Company, 34
poster parks, 41
Poughkeepsie, NY, 35, 156
Priestley, J. B., 131
privitism, 80, 135, 156
Prohibition, 23
prospect and refuge, 122–23
Providence, RI, 111
Pushkarev, Boris, 125–26
Putnam, Robert, 75

railroad crossing signs, 58, 68
railroads, 8, 15, 38, 140
religion, 74, 76, 79, 89–91, 95
Relph, Edward, 126
Roadside Bulletin, 136–37, 153–54
Roadside Business Association, 159, 161
Robinson, John, 132
Roche, John D., Jr., 85
Rochester, NY, 151
Rock City at Chattanooga, TN, 36–37
Ruby Manufacturing Company, 37
Rucker, Evelyn, 102, 104
Ryden, Kent, 89

St. Louis Gunning Advertisement Company v. City of St. Louis, 149
St. Louis, MO, 12–13, 43, 149, 165
St. Paul, MN, 13, 18
Salt Lake City, UT, 64
San Diego, CA, 81–82, 162
San Francisco, CA, 40, 155
sandwich boards, 8
schema/schemata, 119–20
Seattle, WA, 13
Sebree, KY, 24
Seely, Bruce, 55
semiotics, xi–xii, xxix, 7, 49
Sessions, Gordon, 55
Shedd, Peter, 160
shopping, 19, 25–27
shopping centers, 29, 44–45

showcards, 27–28
Sierra Club, 166
sign regulation, xix, 46, 80, 113, 118, 138, 141–42, 144–66; in Alaska, 160; in Hawaii, 160; in Maine, 160; in Rhode Island, 161; in Vermont, 156, 158, 160–61
signlike buildings, xii, xix, 44–45
signs: defined, xvii–xviii, xxi, xv, xxvi, xxix–xxxii, 7; design of, 5, 37–38, 47–49, 98, 117, 127–30, 132–34; digital, 167–68; functions of, xvii, xxxi–xxxii, 31, 117–18, 142, 167
Signs of the Times, xii, 5, 12, 19, 23, 36–37, 45–46, 57, 60, 82, 139, 151, 156–57, 164
small town main streets, xxi, 18–31, 40, 45, 69
snipe signs, 35–36, 40–41, 137, 144
Snow, Todd, 126
social ordering, x, xxi–xxii, xxix, xxxi–xxxii, 26, 73–75, 90, 95–98, 100, 107–08, 167
social stereotyping, xxvii–xxiv, 96–105
South Pass City, WY, 71
Spanish American War, 147
Springfield, IL, 69
Stacey, Judith, 90
Stahlbrodt, Edward A., 151
Steinbeck, John, 83
stop-and-go traffic lights, xxxi–xxxii, 63–65
storefront remodeling, 27–30
storefront signs, xix, xxiii, 3–5, 8, 16, 19–24, 27–30, 123, 130
streetcars, 3, 38, 56–57
Studebaker Corporation, 36
suburbs, xviii, 38, 41, 136
symbolic interactionism, xxi, xxiv, xvii–xxix, 167, 169

Taft, Lorado, 85
taxpayer strips, 39
territoriality, xxiv, xxvi, 95, 107–13, 167
Texas Transportation Institute, 70
Thomas Cusack Company v. The City of Chicago, 150
Tocker, Phillip, 159–60
Tocqueville, Alexis de, 75
Tönnies, Ferdinand, 73–82
Topeka, KS, 146, 148
Traffic Audit Bureau, 33
traffic signs, xix, xxi–xxiii, 52–72, 112, 121, 124, 127–30; in Connecticut, 66, 69; in Illinois, 69; in Maryland, 59–60; in Michigan, 69; in New Jersey, 70; in North Carolina, 66, 69; in Ohio, 60, 69; in Oregon, 70; in Pennsylvania, 69; in Wisconsin, 69
Trenton, NJ, 65
Tunnard, Christopher, 125–26

U.S. Board of Standards, 57
U.S. Bureau of Roads, 60
U.S. Chamber of Commerce, 57
U.S. Constitution, 162, 166
U.S. National Park Service, 71
U.S. Route 66, 83–84

Venturi, Robert, xii, 44
verbalization, xxviii, 7, 36, 98, 106, 119, 133
Vernon, John, ix
visual blight, xix, 7, 10, 16, 34, 50, 62, 118, 135–40, 152
visual culture, xxi, xxxii
visualization, xviii, xxii–xxiii, 3, 117–20, 127, 130, 133, 142–43

Walker and Company, 39
Washington, D. C., 66
way finding, 117, 121–27
Weightman, Barbara, 87
Wilder, Thornton, 73
Williams, Arthur, 14
Williams, George, 5
Willoughby, OH, 35
Wilson, Ruth I., 158

Wilson, William, 146
window displays, xix, 8, 20, 25–31, 42
World War I, 4, 14, 25–27, 81, 151
World War II, 29
Wright, James C., 140
Wrigley Company, 15–16, 38
Wuthrow, Robert, 75

Xerox Corp., 168

Yellowstone Trail, 70
Young, G. H. S., 13

Zelinsky, Wilbur, 77, 86, 167
zoning, 155–59

AMERICAN LAND AND LIFE SERIES

Bachelor Bess: The Homesteading Letters of Elizabeth Corey, 1909–1919
Edited by Philip L. Gerber

Circling Back: Chronicle of a Texas River Valley
By Joe C. Truett

Edge Effects: Notes from an Oregon Forest
By Chris Anderson

Exploring the Beloved Country: Geographic Forays into American Society and Culture
By Wilbur Zelinsky

Father Nature: Fathers as Guides to the Natural World
Edited by Paul S. Piper and Stan Tag

The Follinglo Dog Book: From Milla to Chip the Third
By Peder Gustav Tjernagel

Great Lakes Lumber on the Great Plains: The Laird, Norton Lumber Company in South Dakota
By John N. Vogel

Hard Places: Reading the Landscape of America's Historic Mining Districts
By Richard V. Francaviglia

Landscape with Figures: Scenes of Nature and Culture in New England
By Kent C. Ryden

Living in the Depot: The Two-Story Railroad Station
By H. Roger Grant

Main Street Revisited: Time, Space, and Image Building in Small-Town America
By Richard V. Francaviglia

Mapping American Culture
Edited by Wayne Franklin and Michael C. Steiner

Mapping the Invisible Landscape: Folklore, Writing, and the Sense of Place
By Kent C. Ryden

Mountains of Memory: A Fire Lookout's Life in the River of No Return Wilderness
By Don Scheese

The People's Forests
By Robert Marshall

Pilots' Directions: The Transcontinental Airway and Its History
Edited by William M. Leary

Places of Quiet Beauty: Parks, Preserves, and Environmentalism
By Rebecca Conard

Reflecting a Prairie Town: A Year in Peterson
Text and photographs by Drake Hokanson

A Rural Carpenter's World: The Craft in a Nineteenth-Century New York Township
By Wayne Franklin

Salt Lantern: Traces of an American Family
By William Towner Morgan

Signs in America's Auto Age: Signatures of Landscape and Place
By John A. Jakle and Keith A. Sculle

Thoreau's Sense of Place: Essays in American Environmental Writing
Edited by Richard J. Schneider